Literatures, Cultures, and the Environment

Series Editors
Ursula K. Heise
University of California, Los Angeles
Los Angeles, CA, USA

Gisela Heffes
Johns Hopkins University
Baltimore, USA

Literatures, Cultures, and the Environment focuses on new research in the Environmental Humanities, particularly work with a rhetorical or literary dimension. Books in this series explore how ideas of nature and environmental concerns are expressed in different cultural contexts and at different historical moments. They investigate how cultural assumptions and practices, as well as social structures and institutions, shape conceptions of nature, the natural, species boundaries, uses of plants, animals and natural resources, the human body in its environmental dimensions, environmental health and illness, and relations between nature and technology. In turn, the series makes visible how concepts of nature and forms of environmentalist thought and representation arise from the confluence of a community's ecological and social conditions with its cultural assumptions, perceptions, and institutions.

Rachel Fetherston

Theorising the Postcolonial Eco-Novel

Unsettlement and the Nonhuman in Australian Ecofiction

palgrave
macmillan

Rachel Fetherston
School of Communication and Creative Arts, Faculty of Arts and Education
Deakin University
Burwood, VIC, Australia

ISSN 2946-3157 ISSN 2946-3165 (electronic)
Literatures, Cultures, and the Environment
ISBN 978-3-032-04465-5 ISBN 978-3-032-04466-2 (eBook)
https://doi.org/10.1007/978-3-032-04466-2

Cover illustration: Cover courtesy of Chris McCormack

This Palgrave Macmillan imprint is published by the registered company Springer Nature Switzerland AG.
The registered company address is: Gewerbestrasse 11, 6330 Cham, Switzerland

For Chris, Murphy and Baby B

Acknowledgements

I would like to firstly acknowledge the Traditional Custodians of the different lands on which this research has taken place: the Wurundjeri people and the Djaara people. I pay my respects to Elders past and present, and recognise the incredibly significant contributions of First Nations writers and thinkers to environmental thought and scholarship, both in Australia and around the world.

Much of the research presented in this book stems from a PhD project undertaken at Deakin University on Australian ecofiction, and so would not have happened without the tireless efforts of my supervisors, Emily Potter, Kelly Miller and Devin Bowles. You have provided so much invaluable guidance and I am deeply grateful for your support and vital multidisciplinary perspectives. Thank you to Clare Archer-Lean, Jessica Gildersleeve and Stephanie LeMenager for your exceptionally thoughtful and encompassing feedback on my original thesis, which I have hopefully applied to this book with some success.

My PhD research was undertaken with the support of a Higher Degree by Research Scholarship at Deakin University from 2018 to 2021, with the additional support of a Deakin Post-PhD Publication Scholarship which allowed me to think and write further on some of the key sections of my thesis.

I would also like to thank all of those I connected with at the ASAL 2018 Literary Studies Convention, the Colby College 2019 Summer Institute in the Environmental Humanities, the ASLE Paradise on Fire 2019 Conference, and the ASAL/ASLEC-ANZ Recentring the Region

2023 Conference. Thank you for the opportunity to discuss and workshop some of the research included in this book.

Sincere thanks to Linda Braus at Palgrave Macmillan, the series editors Ursula K. Heise and Gisela Heffes, and the anonymous reviewers of my proposal and manuscript for their invaluable feedback and support.

Thank you to Kathy, Simon, Marcus, Renata, David, Lyn, Mike, Emma, Cathy, Billy, Issy, Georgia and DSC friends for all of your love and support, both during and after the years undertaking the research for this book.

To Chris—you have been with me through all of the highs and lows of this project and have provided so much love and encouragement, particularly when my motivation was waning. Thank you for always supporting me.

Lastly, I want to acknowledge my nonhuman companions who have in many ways led me in this research. My beloved Murphy—you inspire me to do better every day, for both animals and humans. Thanks also to Peck, Monroe, Lamarr, Flynn, Chaplin, Olivier and Bogart. And to all of the other nonhuman beings I've encountered in my life, thank you. You have taught me so much.

Some of the research discussed in Chap. 6 is a revised and extended version of research published earlier in the following journal article:

Fetherston, Rachel. 2021. "Little difference between a carcass and a corpse": Ecological crises, the nonhuman and settler-colonial culpability in Australian crime fiction. *Journal of the Association for the Study of Australian Literature* 21 (2): 1–17.

Competing Interests The author has no competing interests to declare that are relevant to the content of this manuscript.

Praise for *Theorising the Postcolonial Eco-Novel*

"Fetherston offers a bold imagining of recent Australian ecofictions, and their challenge to settler-colonial environmental exploitation and assumptions of belonging. By bridging literary and popular fiction in the context of climate catastrophe, the work contributes to the shape of the ecofiction genre. It will interest scholars in ecocriticism and Australian literature globally, affirming the power of storytelling in the environmental humanities."

–Clare Archer-Lean, *Senior Lecturer in English, University of the Sunshine Coast, chief investigator, ARC-funded Reading Climate project*

"*Theorising the Postcolonial Eco-Novel* is a valuable critical exploration of the significance of contemporary Australian ecofiction. Tracing various manifestations of the non-human other as it appears in these novels, Fetherston offers a compelling analysis of the possibilities of empathising with the other and the ethical potential of ecofiction."

–Jessica Gildersleeve, *Professor of English Literature, University of Southern Queensland*

Contents

CHAPTER 1

Unsettling Ecocriticism

1.1 Ecocriticism: Postcolonial or Postcolonising?

"Humans are the animals that believe the stories they tell about themselves"—so says Mark Rowlands in *The Philosopher and the Wolf* (2009, 2), a quote used by Catherine McKinnon in the epigraph of her 2017 novel, *Storyland*. There is no doubt that storytelling practices have long been of importance to humans across cultures and generations, but the idea of humans as animals telling and believing their own stories reads strangely in the twenty-first century. Humans are indeed animals, but many of us in Western cultures deny and even abhor this fact—and yet it may just be our storytelling that brings us closer to those animals and other nonhumans with whom we share the planet.

Within both the humanities and the social sciences, there is still much to be learnt about the persuasive power of narrative to do just this: persuade readers to be aware of and even empathise with the nonhuman plight. In Australia and other setter-colonial nation states, such knowledge is invaluable given the enduring dominance of Western-colonial ideologies that prioritise instrumentalism and human/nature separation. If Rowlands is right, and we are really to believe the stories we tell, then researchers and communicators may be able to act on such knowledge to enable audiences to empathise with those nonhumans that are so drastically affected by human-induced environmental disaster.

R. Fetherston, *Theorising the Postcolonial Eco-Novel*, Literatures, Cultures, and the Environment,
https://doi.org/10.1007/978-3-032-04466-2_1

In the environmental humanities, the continuing emergence of research that addresses entanglements between humans and nonhumans suggests there is a need amongst scholars to understand the role the nonhuman plays in the human experience, and vice versa. In Australia, native species and habitats are facing immense threats, with the nation's colonial history and its lasting influence within modern Australian culture often blamed for the problematic, instrumentalist attitudes of many white, non-Indigenous Australians towards the environment. The presence of non-native species such as the feral cat is resulting in widespread decimation of native bird and mammal populations (see Doherty et al. 2017), whilst scientists are already mapping the effects of the climate emergency in areas unique to Australia, such as the widely known coral bleaching events on the Great Barrier Reef (see Ainsworth et al. 2016) and the extinction of the Bramble Cay melomys,[1] now considered the first mammal species to become extinct due to anthropogenic climate change (Waller et al. 2017).

Over the past two decades, a steady rise in the publication of environmentally focused fiction in Australia has provided opportunities for readers and researchers to interrogate Australian perspectives on nature and the nonhuman in a literary context. This book reflects not only the need to understand how such fiction represents nature and the nonhuman, but also how the form of the postcolonial novel is specifically positioned to engage with and, at times, challenge Western-colonial notions of belonging and settlement. This work brings together different modes of literary studies practice in what I hope is a productive manner, providing an opportunity to theorise about the power of environmental narratives to effect real-world change and how such knowledge about these narratives might influence the future of hermeneutic ecocritical research.

I am interested in how some Australian novels of the past decade—what I term "postcolonial eco-novels"—provide nuanced portrayals of the nonhuman through their critique of the role that colonisation has played in the climate crisis. More specifically, Australian authors are demonstrating an increased engagement with ecological perspectives outside of dominant settler-colonial views on the nonhuman. Significant to my work here is the idea that such an engagement also differs across literary and genre texts.

[1] The Bramble Cay melomys was a species of rodent native to Australia and found only on Bramble Cay, a coral cay located in the northern part of the Great Barrier Reef.

Both Indigenous and non-Indigenous Australian authors of ecofiction have, over the past decade, established a postcolonising eco-literary framework that connects the concepts of nonhuman agency and more-than-human relationality with the increasingly discussed notion of unsettlement, or unsettled belonging, in the context of climate crisis. Such texts reveal what Kathrin Bartha-Mitchell refers to as "cosmological visions" of the Anthropocene that emphasise "human entanglement with, and emplacement within, an ecosystem" rather than "dualistic constructions of human mastery over 'nature'" (2024, 2). Of course, not all Australian ecofiction novels approach this in the same way; this monograph addresses such differences and positions Australian ecofiction and the "eco-novel" in a transnational context. Graham Huggan argues that "the 'postcoloniality' of national literatures such as Australia's is always effectively *transnational*" (Huggan 2007, viii; emphasis in original). This is because such literatures are "derived from an apprehension of internal fracture (e.g., via the figure of the culturally hyphenated migrant), or from a multiplied awareness of the nation's various engagements with other nations, and with the wider world" (Huggan 2007, viii). The Australian eco-novel is, by extension, equally transnational. The development of the eco-novel form over the past decade speaks to global literary developments in the creative expression of postcolonial and environmental discourses, reflecting an increasing concern amongst fiction writers about the widespread impacts of settler-colonialism, industrialism, extractivism and capitalism. But it is also concerned with how these complexities are navigated "internally", amongst the increasingly diverse peoples and cultures from different parts of the world that now call Australia home.

While this book is certainly concerned with the human as animal, I am just as concerned with the animal, and other nonhumans, as they exist beyond, or "more than" the human, and what literature can reveal to us about them. This may seem counterintuitive—understanding the more-than-human through a creative form that is, of course, human in origin. Nevertheless, while the human animal cannot speak for all animals and all nonhumans, this book argues that attuning ourselves to the more-than-human within our own literary expression is a revealing and generative way of understanding our current ecological moment, and the more-than-human beings who inhabit, affect and are affected by this moment as well. I consider the more-than-human broadly here in the hopes of understanding how unsettlement through ecological crises is evoked beyond the human in literature—what can representations of the nonhuman, be they animals, plants, other living beings, habitats, ecosystems, climate, or the "abiotic nonhuman" such as rocks and products of industry, tell us about the nonhuman, as it exists both separately and in relation to the human? As Tony Hughes-D'Aeth writes, "knowing that once-stable planetary systems are being altered by us changes the way we think and exist, not just in material ways, but in terms of the basic fantasies that hold us as speaking subjects" (2023, 290). Attempting to understand nonhuman ways of being and how we perceive these ways in literature can, I believe, assist the Western human subject in confronting their own role in the Anthropocene while also approaching the idea that the nonhuman asserts their own agency in this epoch.

My exploration of the novels discussed in this book engages with how settler-colonial preconceptions of nature have led to Australia's current moment of ecological crisis, with the aim of challenging these preconceptions and bringing to light how authors both contest and reproduce them in their work. This aligns with a "transcultural" kind of ecocriticism, which prioritises an approach "where Western modes of conquest, categorisation and extraction are checked in order to embrace a multi-vocal array of complex expressions" that go beyond the human (Cooke and Denney 2021, 13). The challenge in approaching contemporary Australian ecofiction through a postcolonial, ecocritical lens is the possibility that broader postcolonial ecocritical theory cannot address the specifics of Australia's socio-political position as a postcolonial nation state, nor its position in a globalised world. This concern stems from Aileen Moreton-Robinson's argument that Australia is not postcolonial "in the same way" as countries such as India and Algeria because Australia's postcoloniality "remains based on whiteness" (2003, 30). Cameo Dalley similarly describes how Australia "remains colonial" and is a place "where reckonings with truth-telling about past injustices remain in their infancy" (2021, 365).

Moreton-Robinson instead uses the term "postcolonizing" to describe contemporary Australian settler society:

> In postcolonizing settler societies, Indigenous people cannot forget the nature of migrancy and we position all non-Indigenous people as migrants and diasporic. Our ontological relationship to land, the ways that country is constitutive of us, and therefore the inalienable nature of our relation to land, marks a radical, indeed incommensurable, difference between us and the non-Indigenous. This ontological relation to land constitutes a subject position that we do not share, and which cannot be shared, with the postco-

lonial subject whose sense of belonging in this place is tied to migrancy. (2003, 26)

One question I address in this book, then, is whether contemporary Australian ecofiction is postcolonial—that is, whether it is interrogating the ways in which people and the more-than-human continue to be embroiled in the ongoing impacts and complexities of settler-colonialism in Australia—or whether the genre is largely postcolonising—that is, consisting of texts that reinforce dominant settler-colonial conceptions of human/nature relations and is therefore involved in the continuing ideological and material colonisation of humans and nonhumans. After all, "the colonials did not go home" (Moreton-Robinson 2003, 26).

Naturally, decolonisation and decoloniality are also relevant here. Ming Dong Gu is careful to distinguish between these two terms, describing decolonisation as "the praxis to remove colonial rule" and decoloniality as "the outcome of decolonization" (2020, 598). With this in mind, Gu asserts that "decoloniality should be concerned with the present condition shaped by the colonial past" (2020, 598). Catherine E. Walsh and Walter D. Mignolo view decoloniality as "born in response to the promises of modernity and the realities of decoloniality" (2018, 4). Significant in the context of this book is Yin Paradies' point that one of the most vital steps in the decolonial process is for all groups of people—both Indigenous and non-Indigenous—to "return to the land, in the understanding that the best way to make amends for colonial pasts is for everyone to mend and make decolonial futures in the present" (2020, 442).

Some of the novels discussed in this book speak to a similar decolonial ethic, particularly those that are situated outside of a white, settler Australian literary tradition. Alexis Wright's *The Swan Book* (2013) and Anita Heiss' *Bila Yarrudhanggalangdhuray* (2021) are both works of Indigenous Australian literature that present long-established understandings of complex human–nonhuman relations within First Nations cultures while condemning the role of the settler-colonial project in the continent's cultural and ecological devastation. Of all of the texts discussed in this book, then, Wright's and Heiss' novels are most actively involved in a decolonising literary practice. As non-white, non-Indigenous Australian authors, Merlinda Bobis' *Locust Girl* (2015) and Yumna Kassab's *Australiana* (2022) also challenge the dominant Anglo-Western storytelling practices of place, belonging and human–nature relations that are typically demonstrated by white settler Australian authors. However, ecofiction novels by white, non-Indigenous authors also have a part to play in condemning and moving away from the settler-colonial exploitation and destruction of Australian nature and the nonhuman. My discussion of these texts recognises a shift within Australian settler-colonial literature towards more nuanced perspectives of nature and the nonhuman that extend beyond current settler-colonial stereotypes of Australian environments, although I do not go as far as suggesting that this kind of storytelling plays an equal role in decolonising literary practices as the work of Wright and Heiss does.

Before I arrive at my specific engagement with these texts, it is important to describe how these novels, and others like them, can be defined as Australian ecofiction and how this relates to the broader global context of ecofiction and other environmental literature. Internationally, contemporary fiction authors are recognising the need to reflect the modern environmental crisis in their works, with a particular emphasis on the climate emergency. Novels such as Claire Vaye Watkins' *Gold Fame Citrus* (2015), Kim Stanley Robinson's *New York 2140* (2017) and Richard Powers' *The Overstory* (2018) as well as older texts such as Octavia Butler's *Parable of the Sower* (1993) and Maggie Gee's *The Ice People* (1998) hint at an international shift towards a fiction that refuses to ignore the modern world's most pressing crisis. Many authors imagine such catastrophe through the depiction of nonhuman experiences alongside those of the human. Speculative novels such as Margaret Atwood's *MaddAddam* trilogy (2003; 2009; 2013), Jeff VanderMeer's *Southern Reach* trilogy (2014) and Paolo Bacigalupi's *The Windup Girl* (2009) grapple with the crises of human-induced habitat destruction, pollution and climate change through portrayals of the nonhuman other in its various forms.

In Australia, trends have been similar. Authors from across the country are increasingly engaging with environmental problems prevalent in both Australia and across the world, with some focusing on the plight and experiences of the nonhuman within the framework of the modern ecological crisis. George Turner's *The Sea and Summer* (1987), an Australian science fiction classic, predates the modern trend of climate fiction, but still engages with themes of the Anthropocene. More recently, Australian literary fiction has demonstrated a strong preoccupation with the idea of anthropogenic environmental destruction, with climate change and discussions of human–nonhuman relationships at the forefront of many of these fictional explorations. Some of these narratives address these issues through strong symbolism and metaphor, such as Charlotte Wood's *The*

Natural Way of Things (2015) and Robbie Arnott's *Flames* (2018), while others like Wright's *The Swan Book* and James Bradley's *Clade* (2015) more overtly convey a sense of urgency regarding climate change in an Australian setting.

This book's engagement with a range of sub-genres—including eco-crime, historical fiction and speculative fiction—under the broad umbrella genre of "ecofiction" and the "postcolonial eco-novel" is particularly important. Integrating considerations of both popular genre and more literary forms provides, I believe, a more nuanced, encompassing understanding of the value of environmental literature that goes beyond the purely "literary" or "high-brow".

Notably, there has been a recent trend within Australian crime fiction to emphasise natural environments (although often with a focus on Anglo-agricultural landscapes) and ecological issues. Novels such as Jane Harper's (2016), and its sequel, *Force of Nature* (2017), Chris Hammer's (2018) and Mark Brandi's *Wimmera* (2017) highlight the similarities between intra-human violence and the violence humankind perpetuates against the natural world. Drought in particular is a common feature in some of these novels. Whilst a recurring theme in Australian cultural expressions over the past century, depictions of drought are also related to the increasing concern amongst scientists, farmers and the general public that anthropogenic climate change will result in more intense and catastrophic drought periods (see Jones et al. 2013; CSIRO and Bureau of Meteorology 2020). Hence, it is not only so-called "serious" literary works that are addressing ecological crises in their narratives, but popular genre as well. I explore the influence of popular genre in more detail in Chap. 2 and then crime fiction specifically in Chap. 6, while also discussing historical fiction in Chap. 7 as a similarly affective genre that is, unsurprisingly, well positioned to comment on the limitations and possibilities of different environmental histories in attending to climate crisis and destructive colonial legacies.

Understanding Australian ecofiction as transnational and global is important here. To discuss this, I must first contextualise it within broader Australian literary traditions. But, as Nicholas Birns argues, Australian literature is far from easy to describe (2015, 1). Part of the difficulty in defining the continent's literature lies in the plurality of cultural identities that exist within its geographical borders. Birns explains how

> [i]n today's globalised world, Australia is not really isolated, and Australian space contains even within its domestic borders many plural national imagi-

> naries stemming from worldwide hybrid and diasporic identities, not to mention the potential permeability of Australia's seacoast, as recent refugee flows have epitomised. (2015, 1)

One might argue that these national imaginaries are what characterise Australian literature and Australian literary criticism as so indefinable, particularly in recent decades when the myth of the "Australian settler-colonial battler" has been challenged and critically repositioned by more diverse perspectives of what it means to be Australian. Jessica Gildersleeve puts forward this notion in her description of Australian literary criticism as a discipline that "disturbs and disrupts idealizations of Australian identity that depend on concepts of the battler, mateship, and masculinity" (2021, 4). Indeed, I would argue that my discussion of some of the novels in this book does just this, as I explore how some Australian settler-colonial authors are challenging more traditional colonial perspectives on nature and the nonhuman in their work.

While Australian literature in the twenty-first century encompasses a wide array of diverse writers and stories, many of the texts analysed in this book address, and are told from the perspectives of, non-Indigenous and white Australians. Borrowing Emily Potter's description, most of the novels here reflect "a generalized, arguably Anglo-Celtic dominant constituency" (2019, 11) in how they perceive the Australian nonhuman and natural world. As I will go on to explain, these novels significantly challenge Western, settler-colonial perceptions of a conquerable and inferior natural world, often depicting the nonhuman through a more ecocentric lens of multispecies relationality.

Ecofiction is in some ways just as complicated as our understandings of Australian literature. Rather than standing alone as an exclusive genre, the term "ecofiction" describes an umbrella genre, encompassing works that follow an environmental theme but that may also exist under other genres such as crime, romance, fantasy, speculative and science fiction, and historical fiction. I view ecofiction novels as works that explore the modern environmental crisis both implicitly and explicitly. Ranging from novels such as Jeff VanderMeer's *Southern Reach* trilogy (2014) and David Mitchell's *The Bone Clocks* (2014) that explore ecological themes from a more symbolic standpoint, to Barbara Kingsolver's *Flight Behaviour* (2012) and N.K. Jemisin's *The Fifth Season* (2015) that feature environmental catastrophe as an obvious plot driver, ecofiction encompasses a wide variety of texts that engage with environmental crises in myriad ways.

In a broader sense, ecofiction constitutes fiction written in the twentieth and twenty-first centuries that reflects or responds in some way to the rapid technological and environmental changes that have occurred globally over the past century. The advent of more contemporary ecofiction could be said to align with, or come soon after, the advent of the modern environmental movement, arguably kick-started by Rachel Carson's *Silent Spring* in 1962. This aligns somewhat with Jim Dwyer's observation that the earliest mention of "ecofiction" occurred in a 1971 fiction anthology (2010, 2). However, there is an argument to be made that ecofiction emerged as a genre earlier in the twentieth century or even the late nineteenth century when "political and literary resistance to industrialism" began to take root (Dwyer 2010, 10). Dwyer broadly defines ecofiction as any fiction that addresses "environmental issues or the relation between humanity and the physical environment, that contrasts traditional and industrial cosmologies, or in which nature or the land has a prominent role" (2010, 2). As I similarly argue in this book, Dwyer considers ecofiction a

> composite subgenre made up of many styles, primarily modernism, postmodernism, realism, and magic realism, and can be found in many genres, primarily mainstream, westerns, mystery, romance, and speculative fiction. (2010, 3)

I must emphasise that "ecofiction" is not simply another term for climate fiction, as it includes texts that may not touch on the climate crisis in an explicit way.[2] Similar to Bartha-Mitchell's "cosmological readings" of contemporary Australian fiction, I understand ecofiction as encompassing works that demonstrate the "potential to wrestle with some aspect of the Anthropocene [but] would not necessarily stand out as 'environmental literature' or climate fiction" (2024, 3). Through character, plot or setting, a work of ecofiction may interrogate a range of issues related to the environment, including but not limited to climate change, species extinction, land management, extractivism and the capitalist-colonial systems that have facilitated such issues to begin with. Importantly, I use the term "ecofiction" rather than "environmental literature" as I would like to

[2] The crime novels discussed in Chap. 6, for example, do engage with environmental change and devastation, such as through bushfire and drought, but do not explicitly refer to climate change.

emphasise the role of fiction specifically in furthering reader understandings of the nonhuman.[3]

The Romantic tradition is often credited as a precursor to ecofiction, at least in the West, given its emphasis on transcendental human experiences in nature, but is distinctly different from the ecofiction we know today. This is because Romanticism "preced[ed] the development of modern ecological science" in the latter part of the nineteenth century while also lacking "the activist quality of much ecofiction" (Dwyer 2010, 9–10). Some of the ecofiction novels discussed in this book are arguably more indebted to non-Western literary traditions than they are to Romanticism, such as the oral storytelling traditions of Australian First Nations groups in the case of Wright's *The Swan Book* (2013) and Arab storytelling practices in the case of Yumna Kassab's *Australiana* (2022). Just as a diverse range of knowledge systems influence contemporary ecocriticism (Cooke and Denney 2021, 14), contemporary ecofiction of the twenty-first century is informed by various multicultural storytelling practices from across the globe.

So if this is ecofiction broadly, what is Australian ecofiction? To understand this, one must first consider contemporary Australian society's cultural and economic stake in the natural world. Australia is known for its unique native wildlife and habitats, not least of all its distinctive deserts, rainforests and coasts which are responsible for a large portion of its tourist appeal.[4] Paradoxically, Australia is also known for its mining, logging and agricultural industries that by definition exploit these same environments for profit. Decades of land clearing for the benefit of these industries have resulted in vast amounts of lost native habitat and species, whilst increasing urbanisation has had similarly devastating consequences (Jackson et al. 2017). The Australian Federal Government's failure over recent years to implement effective, long-term climate change policy has attracted harsh criticism from both Australian (Martin 2019; Flanagan 2020) and international commentators (*Financial Times* Editorial Board 2019; Goodman 2020), especially in the face of longer and more intense bushfire seasons and drought periods.[5]

[3] I consider "environmental literature" to encapsulate both environmental fiction and non-fiction works.

[4] For a discussion of ecotourism in Australia, see Weaver (2008).

[5] For a discussion of the link between bushfire, drought and climate change, see Jones et al. (2013).

Potter describes Australia as "a settler-colonial state with pressing, active and unresolved legacies of colonial violence and dispossession" (2020, 1537) and it is with this, and the above related environmental concerns in mind, that I have identified a need to understand how the expanding genre of Australian ecofiction may affect readerly perspectives on nature and the nonhuman. This is why I have chosen to focus partly on the previously stated "Anglo-Celtic dominant constituency" (Potter 2019, 11) in my list of selected texts, as I believe more complex understandings of how Australian settler-colonial literature represents and actively challenges legacies of settler-colonialism are needed. Settler-colonialist ideologies still have a significant influence on the environmental perspectives of Anglo-Celtic Australians, whose views dominate Australian national discourse. Increasing our understanding of how settler-colonial authors may be shifting these perspectives is therefore vital.

It is fair to state that a restrictive nature-culture dichotomy—that is, the perception that nature and culture are mutually exclusive, and opposing, concepts—is still a structuring influence in contemporary Australia. Marcia Langton argues that "an imperial signification of the dominion of civilisation over nature" (1996, 16) is a prevailing theme within Australian settler-colonial culture, evidenced, for instance, by early tourism advertisements for the country, which depicted vast, so-called wild spaces open for the taking. Langton's remarks can be extended to many works of early and contemporary white, settler-colonial literature in Australia which perpetuate this vision of nature as something to be tamed and exploited. May Gibbs' *Tales of Snugglepot and Cuddlepie* (1918) and James Vance Marshall's *The Children* (1959), for example, incorporate these settler-colonial tropes of opening up and taming colonised country, while still romanticising the bush in both a pastoral and a wilderness sense as somewhere to discover, learn from and seek refuge within. Indigenous people are problematically on the margins of, or erased altogether from these colonial visions.

How these settler-colonial tropes play into Australian ecofiction is a complicated question. Australian ecofiction is a genre yet to be precisely defined, at least by publishing industry standards. Although I have identified this as an "umbrella" genre with other sub-genres situated under it, the genre as a whole still demonstrates particular tropes that a reader might expect (Fetherston et al. 2023, 2). Some of these texts are narrated by nonhumans, although the majority of Australian ecofiction that explores the nonhuman in some way tends to do so from the perspective of humans.

Other common tropes of the Australian ecofiction genre include "the portrayal of the present and future impacts of anthropogenic climate change, environmental activism, environmental exploitation such as logging, mining and farming; the depiction of both native and non-native species of flora and fauna; the depiction of environmental phenomena such as drought, flooding and bushfire; and an exploration of human–nonhuman relationships" (Fetherston et al. 2023, 2), all in the context of Australian places and/or natural environments and ecological issues that are of particular pertinence in Australia. The latter point is, of course, the main reason why I am considering Australian ecofiction as a genre in its own right—because these texts are not simply written by Australian authors but engage with environmental issues that are locally relevant to Australian nature, nonhumans, people and places. A feature of many of these novels, and a primary concern of mine in this book, is the notion of "unsettlement" through ecological crisis and increasing social and cultural opposition to enduring colonial ideologies and systems of authority in Australia.

1.2 (Un)settlement and (Un)belonging

To theorise unsettlement is to navigate the problematics and complexities of settler-colonial belonging in the twenty-first century. Of course, every settler-colonial state is different, but the unsettlement of settler belonging certainly carries a global valence that is relevant beyond Australia. My definition of unsettlement is in line with what Erich Steinman describes as "a framework for understanding the place and relationship of non-Indigenous people to Indigenous decolonization and settler colonialism" (2020, 561). Paradies uses the term "unsettling" in a similar way to how I use it here: "in the sense of disturbing and upsetting, as well as in terms of the potential to shift us from our current places and positions" (2020, 448).

Unsettlement is not simply to *feel* unsettled, but to question what settlement actually consists of and to be confronted with the possibility that, as a settler, one does not truly "belong" anywhere. The passivity of the "un" perhaps does not do justice to what is really a powerful political notion—that the coloniser is not entitled to nor guaranteed a sense of belonging, let alone the more material entitlements made possible through Western-colonial concepts such as property, ownership and enclosure that may facilitate belonging in the first place. Embedded in this is also a choice to be made by the settler-coloniser—to not simply be a passive entity that is confronted with unsettlement, but to actively call into question individual and generational claims on stolen land. Key to this is the idea that a feeling of unsettlement should go beyond the purely ideological and actually result in affective action. Corey Snelgrove et al. discuss opportunities for figuring unsettlement in a more activist framework, reflecting on the possibility of "an ethos of 'unsettled solidarities' that moves across time and space, that is a way of being in the world, a set of ongoing relations" (2014, 25).

The prevention of modernity's current and future wrongs therefore should not be contained to the purely representational, then—a notion which might seem to complicate, even negate, the intention at the heart of this book. However, stories have been and still are key in establishing such claims in the Australian colonial nation state but, as I argue throughout this book, they are also key to their undoing. I understand story as central to what might be deemed the unsettling project, for just as stories have been essential to establishing settler belonging in the first place (including their weaponisation for the purposes of excluding or dominating others (Dalley 2021, 365)), so too can they be used to challenge settler belonging and understand where settler identity stands in the broader context of the modern ecological crisis and evolving human–nature relations. Understanding the unsettlement of colonial belonging is impossible to disentangle from ecological concerns because both colonial and capitalist interests are tied up in the Western narrative of modernity. Paradies describes the "truth about the dangers of modernity" as itself "unsettling" (2020, 439). Telling this truth is something that is not just an ethical responsibility but an absolute necessity in combatting colonialism, climate change, racial injustice and patriarchal oppression.

As a settler descendant myself, my readings of these texts naturally come with an unavoidable bias based on my own subject position. While this book is not an autoethnographic work, I am aware that I am critiquing the institution of settler-colonialism from the "inside", as a privileged, white settler descendant. There exists the vital question of "whether it is possible to be *self-reflexively* critical about the political intent and potency of engaging settler narratives of belonging" (Dalley 2021, 357; my emphasis). I attend further to this complexity of my subject position below.

While I discuss the decolonial ethic behind the work of some of the featured authors, this monograph is not itself a decolonising work in the context of Eve Tuck and Wayne Yang's definition (2012). A true decolonial frame is one that, as Dalley summarises, "must involve some tangible change in the life and circumstances of Indigenous people" (2021, 357).

A discussion of unsettlement is not in itself a decolonising action, then, but must and should be cognisant of the political connotations and possibilities of unsettlement in relation to broader decolonial action. I therefore primarily refer to postcolonialism in the context of interrogating the ongoing legacies of colonialism in Australia, as discussed further in Chap. 2, but I also draw on scholarship in decolonial studies at times, particularly when approaching novels such as *The Swan Book* and *Locust Girl* which are more actively invested in a decolonial ethic.

Unsettlement can, of course, also be considered in broader terms beyond the settler, particularly in the context of climate refugees and displacement. This will be explored in Chap. 3 in my discussion of borders and the animal, although the primary context in which I use it throughout this book is in regard to the possibility of dismantling white, settler-colonial domination of peoples and the environment in the twenty-first century.

1.3 Thinking Through Methodology, Terminology and Positionality

In this book, my close-reading analysis of nine Australian eco-novels investigates the representation of the nonhuman in different contexts, which I will elaborate on shortly, and is particularly attentive to the ability of such fiction to destabilise or reinforce settler claims on land and the nonhuman. I am particularly interested in the literary manifestations of the ecological crises present here in Australia, focusing on the local representation of these issues in connection with the broader climate crisis at a global scale. My selection of novels includes a combination of texts by First Nations authors, non-Indigenous, Anglo-Celtic Australian authors writing within a settler-colonial literary tradition, and non-Indigenous Australian authors whose novels reflect non-white, diasporic literary practices. Australian settlers encompass more than just white Australians, the term also including "a range of non-White identities and minorities who may share in solidarity much in common with Indigenous people" (Dalley 2021, 358), and some of the novels discussed reflect this. I have limited my choice of texts to works that have been published within the last fifteen years. I therefore analyse their immediate relevance to Australia's most recent environmental crises, as well as how the representation of the nonhuman relates to broader concerns surrounding the settler-colonial project in Australia.

From the perspective of many humanities scholars, any research that attempts to understand how nature is represented through storytelling is controversial to begin with, as the concept of "Nature" within a Western framework has been problematised over many years. Drawing on the work of Andrew Biro, Timothy Clark says the following of the varied, sometimes problematic, uses of the term:

> In the modern humanities nature has now become a term pulling in opposite directions at once: "either nature is something to be passionately defended against capitalist predations," or "'nature' can always be revealed as an ideological mask for oppressive social relations." (Biro 2005, 8, cited in Clark 2014, 75)

While the term encapsulates "the totality of the material universe" (Clark 2014, 75) as a real, physical entity, we must be reminded that our interpretations of what is "nature" and "natural" are human ones, drawn from our experiences within particular human cultures. In the case of Western culture, this often means an acceptance of an invented dualism between human and nature; the idea "that the human is radically divided in kind from the rest of creation" (Clark 2014, 77).

For some, "ecology" may be a more apt term as this captures the interconnections between humans and nonhumans, between natural and built environments, between the biotic and the abiotic. Timothy Morton reflects this in their work *Ecology Without Nature* (2009), arguing that "the very idea of 'nature' which so many hold dear will have to wither away in an 'ecological' state of human society" (1). This being said, Morton centres understandings of ecology around human notions of "space and place" (2009, 3), and this is why my discussion of nature and/or ecology centres around scholarship within multispecies studies—an area of research that I believe speaks more to the nonhuman experience than other ecocritical perspectives such as Morton's. I am therefore considering ecology and nature in terms of place but also nonhuman agency and experience. I want to make clear that I am not, as Morton would say about many environmental writings, "putting something called Nature on a pedestal and admiring it from afar" (2009, 5). While I recognise the above concerns in my use of the term, I would like to consider "nature" with a lower case "n", as well as other terms such as "ecology", "natural places", and "natural environment", as concepts that relate to, foster and challenge various notions surrounding the nonhuman—including notions stemming from settler-colonial ideologies—and as concepts that are more-than-human and material, rather than solely projections of the human experience.

The use of the term "nonhuman" is similarly fraught, particularly in the context of animal symbolism as used in literature. For example, nonhuman animals in literature—often Western literatures—are often depicted as "scarcely animals at all but schematic elements in an aesthetic or psychological design" (Scholtmeijer 1993, 259, cited in Donovan 2016, 95). Josephine Donovan comments on the significance of this kind of animal representation in literature and how it has been analysed from ideological standpoints; she argues that new modes of analysis are required to move past what Philip Armstrong describes as "reduc[ing] the animal to a blank screen for the projection of human meaning" (2008, 3, cited in Donovan 2016, 95). My work challenges the solely ideological view of the nonhuman in literature in its exploration of how nonhumans are present in these texts beyond the symbolic or representational.

My use of the term "nonhuman", as with "nature", recognises the problematisation of its usage, particularly the issue of homogenising all nonhuman beings through one, catch-all term.[6] While recognising its limitations, I also bring this term into conversation with multispecies understandings of "more-than-human" worldings, hence I also use the term "more-than-human" in the context of relationality between humans, nonhumans and broader ecosystems. I consider the nonhuman, or nonhuman beings and entities, not as simply physical ones existing in the universe as biological beings, but as entities with "way[s] of being", as Deborah Bird Rose and Thom van Dooren describe (2017, 122). I also consider the nonhuman through the lens of material ecocriticism which considers the idea of "nonhuman agents" as encapsulating "animals, plants, living things generally, or fictional beings, too, as well as matter's active, agentic processes at all scales including the cellular, quantum, and even cosmic levels" (Sullivan 2017, 415).

My references to a "climate crisis" or "climate emergency" throughout this book of course speak to the dire impacts that we are already seeing as a result of anthropogenic climate change. I also consider what Rose describes as

[6] Claire Colebrook (2012) discusses something similar in her critique of multispecies posthumanism, which she argues homogenises the existence and experiences of all living beings in an attempt to challenge an anthropocentric worldview.

> ...a much wider set of entwined events that include the great mass extinction event now in process, the acidification of the oceans, the accumulation of plastic waste, the loss of soils and fertility, the loss of rainforests, and of course the rampant consumption that fuels the work of tearing up and wrecking planet earth. (2013, 208–209)

I therefore primarily refer to the issue of anthropogenic climate change as the "climate emergency" or "climate crisis" as it encompasses a range of ecological crises that are all largely interconnected, with climate change often viewed as the most pressing. The "Anthropocene" is a word that I use sparingly based on existing critique that condemns the term's suggestion that all humans are equally responsible for the epoch as well as its problematic validation of human-centricity (see Malm and Hornborg 2014; Haraway 2015). Donna Haraway views the Anthropocene as "more a boundary event than an epoch" and proposes the term "Chthulucene" to cultivate an awareness of "the dynamic ongoing sym-chtonic forces and powers of which people are a part", encapsulating what is "past, present, and to come" (2015, 160). While this term is applicable to my research, especially because of its focus on both human and nonhuman temporalities and assemblages, I prefer the use of "climate emergency" and "climate crisis" to emphasise the extreme volatility and urgency implied by them, just as the texts analysed in this book emphasise.

When referring to "Indigenous Australian people/peoples" throughout this book, I am referring to "those who identify (or are identified) as the First Peoples of Australia, descended from forebears who precede British colonization" (Potter 2019, 10). My use of the terms "non-Indigenous Australians" and "settler-colonial Australians" refers to, as Potter (2019) describes, "those Australians born or settled here" who do not descend from ancestors who precede this colonisation (10). I also refer to an "Anglo-Celtic dominant constituency" (Potter 2019, 11) of Australian authors writing into Australian ecofiction in the twenty-first century. While Potter uses this framing in reference to dominant voices contributing to "the terms of public debate around non-indigenous belonging in the millennial years" (2019, 11), I believe that this same Anglo-Celtic constituency—as it refers to white settler Australians who primarily descend from English and/or Irish, Scottish and Welsh ancestry—still dominates Australian cultural discourse on connection to place and the environment today. I acknowledge, though, that the term "Anglo-Celtic" does not sufficiently encompass the diverse range of non-Indigenous, settler experiences in Australia, nor is it a term necessarily embraced by all of those who may indeed descend from English, Irish, Scottish and Welsh ancestors, particularly given the vastly different experiences of colonisation between these groups themselves in Australia (an Irish settler immigrating to Australia in the mid-1800s to escape the Great Famine, for example, would likely have experienced very different privileges on their arrival compared to, say, a wealthy English landowner with a large farming homestead).

I use the term here, though, as a helpful means of describing the identity of a large portion of non-Indigenous, white Australians who largely still benefit today from the colonial systems of law and governance that stem from British colonisation—systems which themselves "remain overwhelmingly white and Anglo-Celtic in their composition" (Soutphommasane and Low 2023, 199).[7] I therefore acknowledge that much of my discussion in this book surrounds literature authored by non-Indigenous, white Australians whose views represent a particular Anglo-Celtic Australian experience that largely excludes the experiences of many other non-Indigenous Australians, such as those who identify as Asian-Australian, African-Australian, Middle-Eastern-Australian and many other ethnicities. I also recognise, as Potter (2019) does, that the broad descriptor of "non-Indigenous" carries complicated meanings (10) and, as I am considering works published in the twenty-first century—a time in which "non-Indigenous" can refer to a variety of Australians, and not just white Australians—I must emphasise that its usage can also encompass a broad diversity of perspectives on belonging in Australia that are not restricted to white Australians.

With these definitions in mind, it is important to clarify that as a reader and scholar, I identify myself as a non-Indigenous, Anglo-Celtic Australian. My interpretations of the selected novels as well as my exploration of the relevant knowledge systems and scholarship—including those related to Indigenous Australian concepts, such as Country—are from my

[7] Drawing on the work of Noel Pearson, Tim Soutphommasane and Remy Low describe the Australian national identity as consisting of three elements: "Australia is a country built on an ancient Indigenous heritage (reflecting 60,000 years of Indigenous culture), with British political institutions (reflecting the legacy of British colonization and the historical dominance of white Anglo-Celtic Australia) boasting a multicultural achievement (reflecting the successive waves of mass immigration since the end of the Second World War)" (2023, 197).

perspective as a white, settler-colonial Australian who lives and works on the stolen lands of Australia's First Peoples.

With the exception of Chap. 2, this book's chapters are structured based on the specific "kind" of nonhuman being discussed, as will be made clear. Chapter 2 expands on Australian ecofiction as made up of both popular genre and more "literary" fictions, both of which unsettle perspectives on the nonhuman and associated ecological crises, albeit in different ways. I also continue the discussion begun here regarding the increasing relevance of the "postcolonial eco-novel" as a literary form that considers contemporary ecological concerns in the context of (post)coloniality and notions of belonging and unsettlement. I situate my own definition of the postcolonial eco-novel in the context of existing literary scholarship in ecocriticism, postcolonial literature and Indigenous literature, and broader theory in the environmental humanities such as animal studies, multispecies studies, ecofeminism and environmental justice. Importantly, I discuss the particular significance of ecocritical approaches to popular genre, with an emphasis on crime fiction (explored in Chap. 6) and historical fiction (explored in Chap. 7). This chapter therefore considers the specific ability of popular genre texts within the broader genre of Australian ecofiction to reflect notions of unsettlement through an engagement with the nonhuman.

Chapter 2 also discusses the links between First Nations storytelling practices and the Australian postcolonial eco-novel's approach to the nonhuman, climate, Indigenous sovereignty, and non-Indigenous belonging. This will include a problematisation of the idea that the novel is a supposed silver bullet for environmental communication and climate action, particularly given the form's Western origins. An extension of this discussion is my consideration of the issue of nonhuman symbolism in Western and settler-colonial literature and the ways in which the postcolonial eco-novel both endorses and challenges the instrumentalist use of the nonhuman in literature as purely a representation of human thought and experience.

Chapter 3 begins my exploration of the nonhuman in contemporary Australian ecofiction with a focus on the animal in two works of speculative, literary climate fiction. Specifically, I consider how Waanyi author Alexis Wright and Filipino-Australian author Merlinda Bobis interrogate the notion of borders through their depiction of migration, climate refugees and human–animal communication in speculative, near-future settings. In doing so, both authors also challenge traditional structures and chronologies of the Western novel, further reflecting the notion of unsettlement. Key points of discussion in this chapter include Wright's portrayal of the entangled, mobile relations between protagonist Oblivia (a young Indigenous woman) and the black swan in *The Swan Book* (2013) and the role of Indigenous knowledge systems in combatting both climate crisis and colonial exploitation and domination. The protagonist of Merlinda Bobis' *Locust Girl* (2015) is similarly entangled with the nonhuman animal, although Bobis' novel situates this entanglement more specifically in diasporic perspectives on belonging, conflict and border crossings. Both texts work to challenge Western-colonial perceptions of land ownership, borders and the nonhuman animal, while emphasising the links between climate crisis and the unsettlement of non-Indigenous belonging on stolen Indigenous land.

Chapter 4 undertakes a close reading of two Australian eco-novels that explicitly address the potential future impact of climate change on Australian habitats and communities in specific geographical locations. Mireille Juchau's *The World Without Us* (2015) and Catherine McKinnon's *Storyland* (2017) explore the passing of time in places ravaged by environmental catastrophe and feature what I term "habitat stories"—collaborative stories told through both nonhuman and human experiences that encourage deeper connections between the two, with a particular focus on how different vegetations both enable and challenge white settler-Australian belonging in a time of climate emergency. McKinnon's work depicts the changing habitats and community of a specific coastal location—Lake Illawarra—across many generations. Similarly, Juchau's work portrays the devastating impact of logging and mining economies on a rural rainforest town and its inhabitants. This chapter primarily focuses on the role of habitat, local community and settler-colonial cultural change in addressing the impacts of climate change on the human and nonhuman. Significantly, these novels both challenge and advance problematic settler-colonial perspectives of nature and the nonhuman and thus represent a shift in the way non-Indigenous Australian authors are depicting settler-colonial relations with the nonhuman.

Chapter 5 looks at two novels which I define as part of the Australian regional literary tradition that interrogates the inner lives and conflicts of Australians living in regional and rural areas. Both of these novels are also occupied with the influence of what I term the "abiotic nonhuman". In Jennifer Mills' speculative work *Dyschronia* (2018), this is largely represented by a giant plaster sculpture of a and products of the asphalt industry—the main industry on which the fictional coastal town of Clapstone relies. Through its engagement with the nonhuman, the novel considers the impact of post-industrial shifts on colonial systems of power in regional Australia, presenting an increasing disconnect between the human and nonhuman in such circumstances. Yumna Kassab's *Australiana* (2022) is similarly concerned with the environmental and socio-cultural effects on a regional Australian "everytown" upset by both drought and flooding. Through short, interconnected vignettes, Kassab's novel recognises the role of stones, wool, agricultural equipment and other nonliving objects in human–land relations.

In Chap. 6, I investigate contemporary Australian crime fiction as a largely underexplored representation of the modern environmental crisis, discussing how two settler-authored crime novels portray the troubling relationship between human violence, the settler-colonial decimation of Australian environments, and an increasing disconnect between white Australian settler culture and a stable idea of home. Such a relationship indirectly alludes to the impact of a changing climate on Australian communities and ecosystems and suggests that popular genre fiction can contribute in profound ways to broader socio-environmental considerations. The nonhuman focus of this chapter is what I term the "nonhuman phenomena" of drought and bushfire in Jane Harper's (2016) and Chris Hammer's *Scrublands* (2018), and I look at what these novels reveal about the criminal nature of anthropogenic climate change and the settler-colonial destruction of Australian habitats.

Chapter 7, the final close-reading chapter, looks at the genre of historical fiction in the context of Australian postcolonial ecofiction. Specifically, I discuss the representation of water agency and rivers in Anita Heiss' *Bila Yarrudhanggalangdhuray* (2021), and the links between such and Heiss' assertions of Indigenous sovereignty and belonging throughout the novel. This chapter begins with a discussion of the shifting nature of human–water relations globally and the attentiveness to this issue in ecofiction. I recognise Heiss' subversion of the traditional colonial historical novel as an unsettlement in itself, while the events and themes of her novel—which is a fictional account of the real 1852 flood of Gundagai in New South Wales—reveals the inadequacy of settler/land relations in understanding the power of water and rivers. The positioning of this novel in the final close-reading chapter leads into the conclusion's discussion of the risks of exploiting First Nations knowledges and cultural expression for the purposes of non-Indigenous literary and ecocritical scholarship.

My concluding chapter considers whether shifts in nonhuman representation and the increased emphasis on unsettlement in the Australian postcolonial eco-novel reflect the possibilities of a more relational, less instrumentalist ecological future for postcolonial Australia, or whether such possibilities are made miniscule in the face of still-widespread settler-colonial policy and cultural domination in Australia and elsewhere. This final chapter will explore this question as well as the position of the postcolonial eco-novel in a global literary framework and how the form may develop going forward, particularly with decolonising literary practices in mind.

References

Ainsworth, Tracy D., Scott F. Heron, Juan Carlos Ortiz, Peter J. Mumby, Alana Grech, C. Daisie Ogawa, Mark Eakin, and William Leggat. 2016. Climate Change Disables Coral Bleaching Protection on the Great Barrier Reef. *Science* 352 (6283): 338–342.

Armstrong, Philip. 2008. *What Animals Mean in the Fiction of Modernity*. London: Routledge.

Arnott, Robbie. 2018. *Flames*. Melbourne: Text Publishing.

Atwood, Margaret. 2003. *Oryx and Crake*. London: Bloomsbury.

Atwood, Margaret. 2009. *The Year of the Flood*. New York: Doubleday.

Atwood, Margaret. 2013. *MaddAddam*. New York: Doubleday.

Bacigalupi, Paolo. 2009. *The Windup Girl*. San Francisco: Night Shade Books.

Bartha-Mitchell, Kathrin. 2024. *Cosmological Readings of Contemporary Australian Literature: Unsettling the Anthropocene*. London: Routledge.

Birns, Nicholas. 2015. Is Australian Literature Global Enough? *Journal of the Association for the Study of Australian Literature* 15 (3): 1–4.

Biro, Andrew. 2005. *Denaturalizing Ecological Politics: Alienation from Nature from Rousseau to the Frankfurt School and Beyond*. Toronto: University of Toronto Press.

Bradley, James. 2015. *Clade*. Melbourne: Hamish Hamilton.

Brandi, Mark. 2017. *Wimmera*. Sydney: Hachette Australia.

Butler, Octavia. 1993. *Parable of the Sower*. New York: Four Walls Eight Windows.

Carson, Rachel. 1962. *Silent Spring*. Boston: Houghton Mifflin.

Clark, Timothy. 2014. Nature, Post Nature. In *The Cambridge Companion to Literature and the Environment*, ed. Louise Westling, 75–89. New York: Cambridge University Press.

Colebrook, Claire. 2012. Not Symbiosis, Not Now: Why Anthropogenic Change Is Not Really Human. *Oxford Literary Review* 34 (2): 185–209.

Cooke, Stuart, and Peter Denney. 2021. Thinking About Transcultural Ecocriticism: Space, Scale and Translation. In *Transcultural Ecocriticism: Global, Romantic and Decolonial Perspectives*, ed. Stuart Cooke and Peter Denney, 13–31. London: Bloomsbury Academic.
CSIRO and Bureau of Meteorology. 2020. State of the Climate 2020. *Clayton*. http://www.bom.gov.au/state-of-the-climate/documents/State-of-the-Climate-2020.pdf
Dalley, Cameo. 2021. Becoming a Settler Descendant: Critical Engagements With Inherited Family Narratives of Indigeneity, Agriculture and Land in a (Post) Colonial Context. *Life Writing* 18 (3): 355–370.
Doherty, Tim S., Chris R. Dickman, Chris N. Johnson, Sarah M. Legge, Euan G. Ritchie, and John C. Z. Woinarski. 2017. Impacts and Management of Feral Cats *Felis Catus* in Australia. *Mammal Review* 47 (2): 83–97.
Donovan, Josephine. 2016. *The Aesthetics of Care: On the Literary Treatment of Animals*. New York: Bloomsbury.
Dwyer, Jim. 2010. *Where the Wild Books Are: A Field Guide to Ecofiction*. Reno: University of Nevada Press.
Fetherston, Rachel, Emily Potter, Kelly Miller, and Devin Bowles. 2023. Seeking Greener Pages: An Analysis of Reader Response to Australian Eco-Crime Fiction. *Australian Humanities Review* 71 (May): 1–21.
Financial Times Editorial Board. 2019. Australia's Bushfires Have Exposed Its Leaders' Failings. *The Financial Times*, December 23. https://www.ft.com/content/5082d45c-232e-11ea-b8a1-584213ee7b2b. Accessed 27 Jan 2025.
Flanagan, Richard. 2020. 'Australia Is Committing Climate Suicide.' *The New York Times*, January 3. https://www.nytimes.com/2020/01/03/opinion/australia-fires-climate-change.html. Accessed 27 Jan 2025.
Gee, Maggie. 1998. *The Ice People*. London: Richard Cohen Books.
Gibbs, May. 1918. *Tales of Snugglepot and Cuddlepie*. Sydney: Angus & Robertson.
Gildersleeve, Jessica. 2021. Introduction: Australian Literature, Companionship, and Viral Responsibility. In *The Routledge Companion to Australian Literature*, ed. Jessica Gildersleeve, 1–6. New York and London: Routledge.
Goodman, Jack. 2020. What Is Australia Doing to Tackle Climate Change? *BBC News*, January 3. https://www.bbc.com/news/world-australia-50869565. Accessed 27 Jan 2025.
Gu, Ming Dong. 2020. What is "Decoloniality"? A Postcolonial Critique. *Postcolonial Studies* 23 (4): 596–600.
Hammer, Chris. 2018. *Scrublands*. Crows Nest: Allen & Unwin.
Haraway, Donna. 2015. Anthropocene, Capitalocene, Plantationocene, Chthulucene: Making Kin. *Environmental Humanities* 6 (1): 159–165.
Harper, Jane. 2016. *The Dry*. Sydney: Pan Macmillan Australia.
Harper, Jane. 2017. *Force of Nature*. Sydney: Pan Macmillan Australia.

Huggan, Graham. 2007. *Australian Literature: Postcolonialism, Racism, Transnationalism*. Oxford: Oxford University Press.

Hughes-D'Aeth, Tony. 2023. Australian Fiction in the Anthropocene. In *The Cambridge Companion to the Australian Novel*, ed. Nicholas Birns and Louis Klee, 289–304. Cambridge: Cambridge University Press.

Jackson, William, Robert Argen, Nic Bax, Graeme Clark, Sarah Coleman, Ian Cresswell, Kathryn Emmerson, Karen Evans, Mark Hibberd, Emma Johnston, Melita Keywood, Andrew Klekociuk, Richard Mackay, Dan Metcalfe, Helen Murphy, Alex Rankin, David C. Smith, and Barbara Wienecke. 2017. *Australia State of the Environment 2016*. Canberra: Australian Government Department of the Environment and Energy.

Jemisin, N. K. 2015. *The Fifth Season*. London: Orbit.

Jones, Roger, Celeste Young, John Handmer, Adrian Keating, Gayathri Mekala, and Peter Sheehan. 2013. *Valuing Adaptation Under Rapid Change*. Gold Coast, QLD: National Climate Change Adaptation Research Facility.

Kingsolver, Barbara. 2012. *Flight Behaviour*. London: Faber.

Langton, Marcia. 1996. What Do We Mean by Wilderness?: Wilderness and Terra Nullius in Australian Art [Address to The Sydney Institute on 12 October 1995]. *The Sydney Papers* 8 (1): 10–31.

Malm, Andreas, and Alf Hornborg. 2014. The Geology of Mankind? A Critique of the Anthropocene Narrative. *The Anthropocene Review* 1 (1): 62–69.

Marshall, James Vance. 1959. *The Children*. London: Michael Joseph.

Martin, Sarah. 2019. Australia Ranked Worst of 57 Countries on Climate Change Policy. *Guardian Australia*, December 11. https://www.theguardian.com/environment/2019/dec/11/australia-ranked-worst-of-57-countries-on-climate-change-policy. Accessed 27 Jan 2025.

McKinnon, Catherine. 2017. *Storyland*. Sydney: HarperCollins Publishers.

Mitchell, David. 2014. *The Bone Clocks*. London: Sceptre.

Moreton-Robinson, Aileen. 2003. I Still Call Australia Home: Indigenous Belonging and Place in a White Postcolonizing Society. In *Uprootings/Regroundings: Questions of Home and Migration*, ed. Sara Ahmed, Claudia Castada, Anne-Marie Fortier, and Mimi Sheller, 20–37. Oxon and New York: Routledge.

Morton, Timothy. 2009 [2007]. *Ecology Without Nature: Rethinking Environmental Aesthetics*. Cambridge, Massachusetts & London, England: Harvard University Press.

Paradies, Yin. 2020. Unsettling Truths: Modernity, (De-)Coloniality and Indigenous Futures. *Postcolonial Studies* 23 (4): 438–456.

Potter, Emily. 2020. Contesting Imaginaries in the Australian City: Urban Planning, Public Storytelling and the Implications for Climate Change. *Urban Studies* 57 (7): 1536–1552.

Potter, Emily. 2019. *Writing Belonging at the Millennium: Notes from the Field on Settler-Colonial Place*. Bristol: Intellect.
Powers, Richard. 2018. *The Overstory*. New York: W.W. Norton & Company.
Robinson, Kim Stanley. 2017. *New York 2140*. New York: Orbit.
Rose, Deborah Bird. 2013. Anthropocene Noir. *Arena Journal* 41:208–209.
Rose, Deborah Bird, and Thom van Dooren. 2017. Encountering a More-than-Human World: Ethos and the Arts of Witness. In *The Routledge Companion to the Environmental Humanities*, ed. Ursula K. Heise, Jon Christensen, and Michelle Niemann, 120–128. London and New York: Routledge.
Rowlands, Mark. 2009. *The Philosopher and the Wolf*. London: Granta.
Scholtmeijer, Marian. 1993. *Animal Victims in Modern Fiction: From Sanctity to Sacrifice*. Toronto: University of Toronto Press.
Snelgrove, Corey, Rita Kaur Dhamoon, and Jeff Corntassel. 2014. Unsettling Settler Colonialism: The Discourse and Politics of Settlers, and Solidarity with Indigenous Nations. *Decolonization: Indigeneity, Education & Society* 3 (2): 1–32.
Soutphommasane, Tim, and Remy Low. 2023. Tomorrow's Australia: Race and Racialisation in Australian Education. In *World Yearbook of Education 2023: Racialisation and Educational Inequality in Global Perspective*, ed. Janelle Scott and Monisha Baja, 178–206. London and New York: Routledge.
Steinman, Erich. 2020. Unsettling as Agency: Unsettling Settler Colonialism Where You Are. *Settler Colonial Studies* 10 (4): 558–575.
Sullivan, Heather I. 2017. Material Ecocriticism and the Petro-Text. In *The Routledge Companion to the Environmental Humanities*, ed. Ursula K. Heise, Jon Christensen, and Michelle Niemann, 414–423. London and New York: Routledge.
Turner, George. 1987. *The Sea and Summer*. London: Faber.
VanderMeer, Jeff. 2014. *Southern Reach*. New York: Farrar, Straus and Giroux.
Waller, Natalie L., Ian C. Gynther, Alastair B. Freeman, Tyrone H. Lavery, and Luke K.-P. Leung. 2017. The Bramble Cay Melomys *Melomys rubicola* (Rodentia:Muridae): A First Mammalian Extinction Caused by Human-Induced Climate Change? *Wildlife Research* 44 (1): 9–21.
Walsh, Catherine E., and Walter D. Mignolo. 2018. *On Decoloniality: Concepts, Analytics, Praxis*. Durham: Duke University Press.
Watkins, Claire Vaye. 2015. *Gold Fame Citrus*. New York: Riverhead Books.
Weaver, David. 2008. *Ecotourism*. Milton: Wiley.
Wood, Charlotte. 2015. *The Natural Way of Things*. Crows Nest: Allen & Unwin.
Wright, Alexis. 2013. *The Swan Book*. Artarmon: Giramondo Publishing Company.

CHAPTER 2

The Nonhuman and the Postcolonial Eco-novel

2.1 Postcoloniality, the Nonhuman and the Novel

Colonialism, capitalism and the climate crisis are intimately connected—this much we know (Whyte 2017; Haraway 2015a). The same perspective can be applied to any text engaged with environmental themes or discourse—such texts are equally tied to colonial and capitalist ideologies, either in their reinforcement of such, their rejection of such, or both. While I am far from the first literary studies scholar to argue this (e.g., Clark 2014; Potter 2019; Cooke and Denney 2021), one of the main aims of this book is to more explicitly conceptualise a form of novel that attends to the complexities of the enduring impacts of colonialism in contemporary times, and human–nature relations. This is the "postcolonial eco-novel".

As will be demonstrated via my engagement with a range of both literary and genre texts in this book, the borders that surround the postcolonial eco-novel are far from distinct. Key to this conceptualisation in an Australian context is understanding the origins of Australian setter literature, the tropes and archetypes of which are now being challenged by some authors of Australian ecofiction. Written by non-Indigenous authors, Australian settler literature is primarily concerned with interrogating and negotiating settler-colonial understandings of belonging within Australian

R. Fetherston, *Theorising the Postcolonial Eco-Novel*, Literatures, Cultures, and the Environment,
https://doi.org/10.1007/978-3-032-04466-2_2

settings—settings that are often, although not always, heavily stereotyped. The Australian outback and the Australian bush play a central role in many of these narratives, demonstrating the ways in which settler-colonisers seek a sense of belonging in what they perceive to be harsh and unfamiliar natural places. In *Talking to My Country* (2016), journalist Stan Grant comments on the proprietary attitude of settler-colonisers towards land:

I have always thought settler communities unique, they see new lands as something to be claimed, as places without a past. The people who have come [to Australia] have left history behind. Settlers look out; they look forward. The mark of success is to leave a better life for your children. Settlers prove how much they belong by how far they have travelled, how much they have acquired. (32)

Jeanine Leane argues that, following Australia's Federation in 1901, the need to "manufacture through literature a unique and distinct Australian identity" (2010, 32) became widespread. This identity is described by Leane as "a handsome, care free, anti-intellectual bushman, apathetic and unaware of the wider world around him and who embraced and practiced the art of mateship" (2010, 32). The prioritisation of this Australian settler identity resulted in the exclusion or stereotyping of "women, urban people, immigrants and Indigenous people" (Leane 2010, 32).

One question I attempt to answer across this book is whether recently published Australian ecofiction has found ways to move past the colonial desire to control and exploit the nonhuman and the natural world through Cartesian understandings of nature versus culture, and human versus animal. Are any of the texts discussed in this research *decolonising* the way Australians think about humans and nature? Perhaps. Do some of these works simultaneously challenge and utilise traditional elements of Australian settler-colonial literature? Most definitely. This will be a key point of discussion in Chaps. 4–6 as I analyse the ways in which Australian ecofiction both confronts and adheres to some of the long-accepted stereotypes of the Australian natural world as promoted through the settler-colonial agenda—an agenda that has prioritised, and still prioritises, the exploitation and subsequent degradation of Australian nature and the nonhuman in order to reap economic benefits and establish colonisers as the tamers of a supposedly brutal land.

Before continuing, though, I must first establish the ways in which I am using the term "postcolonial" in an Australian context. The Australia that we know today is a nation that has been established through a violent colonial past. Many of its politics, institutions and industries originate in and still perpetuate the colonial systems that have been in place since the First Fleet reached Australian shores in 1788. "Colonialism" can thus be described as "a 'structured relationship of domination and subordination'... [that] establishes and maintains racialized hierarchies of power/knowledge that legitimate, serve, and naturalize the interests of the dominant group" (Barrera 1979, cited in Biermann 2011, 388)—in this case, Anglo-Celtic Australians. With this in mind, I agree with Potter's use of the term "postcolonial" as "the ongoing, continually enacted nature of colonization in contemporary Australia", and not as signifying "an end to the colonial process" (2019, 10). Potter also describes how the term was, in Australia's millennial years, considered "normative nomenclature in critical discussions over the terrain of non-indigenous belonging" (2019, 11). Today, Potter argues, it "still activates a tacit understanding of its meaning as referring to a state of still-negotiated settler-colonial and colonized relations" (2019, 11). Key to this is the overturning of *terra nullius*—"the key imperialist discourse that supported the colonization of Australia ... the idea that the land belonged to no one before the British arrived in 1788" (Rodoreda 2023, 423)—by the High Court of Australia in *Mabo and Others v Queensland (No. 2) of 1992.* This has become colloquially known as the "*Mabo* decision" and constituted a formal recognition of "native title rights to land, rights that should have been recognised upon settlement and that might still be held to exist, under certain conditions, across Australia' (Rodoreda 2023, 423). Australian fiction written and published after this time is therefore often framed as "post-*Mabo* fiction", pointing to how literary representations of history, belonging and land rights changed dramatically in the wake of this landmark ruling. Geoff Rodoreda claims that it is this kind of fiction which is "the key postcolonial current in Australian novel writing today" (2023, 424). Any reference made to "postcolonialism" or "the postcolonial" in my research engages with the significant effect the *Mabo* decision had on the Australian settler psyche, and is therefore not used to deny the still active colonial systems that oppress both Indigenous Australian peoples, and non-Indigenous people who do not fit an Anglo-Australian stereotype as endorsed by the Anglo-colonial project.

I also do not wish to ignore the very real moves by various groups across Australia—particularly Indigenous Australians—to enact decolonisation. This drive for decolonisation in a settler state like Australia is particularly complex, as elucidated by Soenke Biermann:

Caught in the middle of a 'double-bind' of external and internal colonial relationships, the state is formally independent of the original colonial power. However, internally these nation-states are a case of colonists that never left and, for Indigenous peoples, colonialism that never ended. The settlers brought and developed traditions, systems, and structures which have since become the basis of the independent state. Meanwhile Indigenous peoples have been relegated to minority positions in their own countries… (2011, 393)

As Biermann goes on to say, the push for decolonisation in a nation such as Australia is largely focused on "continuing assertions of Indigenous sovereignty and rights, the reclamation of Indigenous philosophies and knowledges as well as the exposition and dismantling of hidden structures, ongoing processes, and unquestioned traditions of colonial governance" (2011, 393). This means that decolonisation involves not just the dismantling of colonial structures and institutions, but also a "re-consideration of the totality of relationships to other people and the land we live in" (Biermann 2011, 393). As discussed in Chap. 1, not all of the novels in this book can be read as a clear call to action for Australia's decolonisation. Nevertheless, there are some elements of these texts that strongly critique not just the structures of colonialism, but colonial ways of being and knowing when it comes to nature and the nonhuman—especially those settler-colonial perspectives that have resulted in large-scale environmental degradation in Australia. In this sense, decolonisation is a relevant theme within my textual analysis.

Perhaps unsurprisingly, then, Indigenous epistemologies factor into some of my analysis of the selected texts, most prominently *The Swan Book* which depicts a continuation of Indigenous dispossession and genocide in a near-future Australia and positions Indigenous presence, culture, and reality as vital parts in the fight for decolonisation. As the critique of settler-colonial perspectives of nature and the nonhuman are central to this book, I recognise that literary practice has a significant role to play in decolonisation, such as through the complex representation of Indigenous voices and lived experiences in literary narratives,[1] but that literary engagement with decolonial ideologies is not the equivalent of the actual, real-world work of decolonisation.

[1] Justine Seran touches on this in her analysis of Sally Morgan's *My Place* (1987), arguing that "the complexity of representation in literary practice functions as a challenge to colonial boundaries" (2015, 664).

Some of the settler-authored texts discussed incorporate *and* challenge some of the norms within the Australian settler tradition. Their focus on the nonhuman as well as human stories in response to the climate crisis may indicate the act of "learning to die" (LeMenager 2017, 229) when faced with the destruction of climate change. Describing climate fiction writers within the Euro-American tradition, LeMenager suggests that "learning to die implies letting go of some cherished tenet of humanism" (229). Such an action on the part of writers may therefore result in "recognizing oneself as a biological entity, challenging traditional humanist ideas of the sovereign self" (229).

A recurring theme in various climate fiction narratives of the past decade, Australian or otherwise, has been that of time and its relationship to human memory. Texts such as David Mitchell's *The Bone Clocks* (2014), James Bradley's *Clade* (2015), Briohny Doyle's *The Island Will Sink* (2016) and Amitav Ghosh's

Gun Island (2019) illuminate the connections between climate change, memory and time—both time in a human sense and the more challenging concept of deep time. Understanding the ways that Australian texts in particular explore these connections is significant given the "young" status often granted to this recently established Western nation, and the juxtaposing ancient timelines of Indigenous Australian peoples living on this continent and forming complex relationships with the climate and the nonhuman.

Potter describes the concept of time as understood through Indigenous worldviews as "an embodied practice that is marked and lived through associations between ecological events and ancestral connections" (2015, 102). It is perhaps such a notion of time that some authors discussed in this book are attempting to capture as they explore the potential impacts of climate change on Australian settler culture. Such explorations of time also set these texts apart from other, earlier works of Australian settler literature that more traditionally utilise a linear Western temporality to explore the settler-colonial relationship with the Australian nonhuman. Potter describes such a temporality as "junctural, marked out by pre- and post-temporality" (2019, 6). Potter also comments that

Time is thus marked in embodied ways; in the travel to place, in gatherings, and in the environmental happenings that signify time's enactment. (2015, 109)

The Indigenous Australian concept of Dreaming—described by Irene Watson as "an ever-present place of before, now, and the future, a place that we are constantly returned to" (2009, 36)—is also a means through which Indigenous understandings of time can be considered. As Potter explains, "Dreaming is distinct from ordinary time.... The former is enduring, while the latter is temporary, subject to constant change" (2015, 103). Such understandings are certainly relevant to the way that some Australian ecofiction authors address concepts of nonhuman time, as well as time in terms of climate change.

2.2 The Eco-novel and Popular Genre

It is important to clarify that Australian ecofiction, as it is considered in this book, encompasses both literary and genre texts and therefore has the capacity to influence its readers in different ways due to variations in accessibility. Within the constraints of this book, I am considering the Australian crime fiction texts as representative of a more popular and less literary genre that sits within the broader genre of Australian ecofiction. I have chosen to examine crime fiction instead of other genres because I see some of the Australian texts in this genre as responding strongly to current ecological crises and the links between such crises and postcolonial violence in Australia. As I will go on to discuss, the violence coupled with the often-quintessential Australian outback settings depicted in these texts provides an opportunity to explore how Australian ecofiction is addressing settler-colonial violence and environmental exploitation in the current moment. With the popularity of a text like Jane Harper's 2016) (as explored in Chap. 6), it is crucial to consider how the recently revived market of Australian crime may relate to a heightened awareness of the climate crisis amongst both authors and readers.

It is also worth noting Ken Gelder's comment, made in 2001, that

> [t]he readerships ... for Australian crime fiction and science fiction and fantasy—as well as adventure thrillers, eco-romances, horror fiction, and so on—have ... barely figured in Australian literary analysis. (118)

The situation has changed since the early 2000s with more scholars since focusing on the merits of Australian crime fiction within the Australian literary space,[2] but the idea of crime fiction as a potential vessel for ideas pertaining to climate change and the nonhuman remains largely underexplored, although emerging scholarship surrounding this topic

[2] Stephen Knight is one scholar who has written extensively on the cultural influence of Australian crime fiction, both in Australia and internationally.

does exist (see Walton and Walton 2018; Rugg 2017; Mrozewicz 2020). As will be demonstrated in Chap. 6, there is potential for readers to connect deeply with natural places through Australian crime fiction. The modern Australian crime texts included in this research are largely concerned with what Gelder and Rachael Weaver describe as "anxieties that flow from a sense of the colonial landscape as a fluid, transient place" (2008, 4), but in more contemporaneous ways; their rural settings, threatened by both drought and flooding, strongly follow this conception of colonial landscapes as unfixed, ever-changing environments. They are therefore not amongst the texts in this study that actively challenge the settler-colonial literary tradition, as described above, but their strong engagement with environmental phenomena in an Australian setting means that they can be identified within the genre of ecofiction.[3]

The advantage of investigating popular genre texts here is the ability to analyse particular tropes and archetypes in relation to how they uphold or challenge certain attitudes towards nature and the more-than-human. This is an expanding field of inquiry outside of crime fiction as well, and so we might look to other popular genre scholarship to see the benefits of this kind of literary research. Science fiction scholar Andrew Milner is well known for his work on "cli-fi" and argues that "Science fiction … is a primary mechanism—perhaps the primary mechanism—by which our culture imagines its possible futures, both positive and negative", including those futures affected by climate change (2020, 77). In other genre realms, Lauren O'Mahony identifies natural environments as a significant feature in the romance fiction of Australian author Marie Bjelke-Petersen (2020, 79). O'Mahony has also analysed death in Australian rural romance with a focus beyond the human, looking at the death of animals, environments and inanimate objects (which I classify as the "abiotic nonhuman" in this book) in this sub-genre of romance fiction (2017). In the fantasy space, Reyyan Bal discusses the largely unrecognised environmental imaginaries put forward by many fantasy authors whose work often "disrupts

[3] It should be noted that the reference to such phenomena as "environmental" or "natural" does not appropriately recognise the connection between such events in Australia (and elsewhere) and anthropogenic climate change. Both severe drought and intensifying fire seasons are impacting rural and urban communities across the continent, with anthropogenic climate change the leading cause of the disastrous conditions currently being faced by many Australians. Such phenomena are therefore environmental, but often strongly influenced by human action. For the purposes of this research, these events will be deemed "environmental phenomena" but with this anthropogenic understanding in mind (Jones et al. 2013, 2).

the reason/nature binary" of Western modernity (2021, 1289). There is clearly impetus, then, to further investigate the potential of crime fiction, as a popular genre, to represent environmental themes and concerns as well.

I also address the role of historical fiction in furthering environmental discourse in Australia, with a particular emphasis on an example of First Nations historical fiction in Chap. 7. In a similar way to crime fiction, there is potential for recognisable features of Australian history to engage readers, particularly when those dominant histories which may be taken as fact by many readers are challenged or complicated by marginalised histories of colonisation and ecological devastation. Greg Forter argues that the historical novel must be "climate-fictional" if it is to adequately address the challenges and complexities of our current moment (2023, 1666). Forter goes as far to say that most historical novels written today are not "authentically historical" because they fail to confront "the constraints on historical thinking imposed by our present" (2023, 1667).

It is also important here to distinguish between the form of the postcolonial eco-novel and that of climate fiction, particularly of the more popular, speculative kind. As the majority of the texts analysed in this book concern climate change in some shape or form, it is necessary to clarify how I approach climate change and climate fiction (or cli-fi[4]) in the context of Australian ecofiction and the postcolonial eco-novel. When I speak of climate fiction, I am referring to a genre of fiction that is "explicitly focused on climate change" (Schneider-Mayerson 2018, 473). According to Tony Hughes-D'Aeth, cli-fi texts are usually dystopian in nature, emphasising the ways in which "the effects of global warming … have produced a catastrophic reorganisation of human society" (2023, 290). These texts can be literary in nature or popular genre—a perspective which is at odds with Milner and J.R. Burgmann's strong assertion that climate fiction should be categorised as a subgenre of science fiction. This is because, they contend, climate fiction has primarily been authored by people who tend to identify as science fiction writers and whose works are often largely concerned with science and technology (2018, 5). While there are certainly many climate fiction texts that fit this mould, I tend to disagree with the premise as it largely ignores the burgeoning interest in climate change in other popular genres—as some of the popular genre

[4] The coining of the term "cli-fi" has been claimed by activist and blogger Dan Bloom (Forthomme 2014).

research mentioned previously contends. It is particularly difficult to ignore the undeniable influence of N.K. Jemisin's critically acclaimed *Broken Earth* trilogy (2015–2017) as a fantasy series deeply engaged with the socio-political and environmental impacts of climate crisis, as well as Jeff VanderMeer's new weird fiction, such as the *Southern Reach* trilogy (2014), which depicts both the horror and the possibilities of human–nonhuman entanglements.

While in the past 10–15 years the Australian literary scene has featured a strong focus on climate fiction, there was previously widespread criticism of Australian literature for its lack of engagement with ecological crises. This was the case for literature worldwide as well; Amitav Ghosh has famously argued for a serious literature of climate change—one that is broadly recognised by writers, reviewers and scholars rather than "relegate[d] … to the genre of science fiction"' (2016, 7). In the case of Australian literature, the early 2000s saw criticisms aimed at literary fiction authors for not engaging enough with ecological discourse in their work. Potter describes how Australian fiction was criticised by various cultural commentators for a lack of "ecological consciousness" (2005, np); in particular, writer Drusilla Modjeska emphasised that it was "failing to attend to the 'conditions on the ground' in Australia, to 'the stuff of lives lived in this particular experience of global modernity'" (2002, 205 and 209, cited in Potter 2005, np). This is particularly shocking given that Australia has largely failed to take on the challenge of climate change (Milner 2020, 77).

In recent years, though, Australian readers have seen more and more Australian literary fiction with a focus on climate change, with many texts, such as Alexis Wright's *The Swan Book* (2013) and Catherine McKinnon's *Storyland* (2017), being recognised by some of Australia's most prestigious literary awards.[5] According to Matthew Schneider-Mayerson, one reason for the focus given to climate fiction worldwide over the past decade is that there is

> the hope that these plays, novels, short stories, poems, and children's stories might lead to a wider and deeper climate consciousness and thereby contribute to more progressive environmental policies and politics. (2018, 474)

This particular assertion is the reason why an increasing number of humanities scholars are utilising more empirical methods of research to

[5] Both Wright's *The Swan Book* and McKinnon's *Storyland* were shortlisted for the Miles Franklin Award in their respective years.

understand whether such a hope is well founded (Schneider-Mayerson et al. 2023; Malecki et al. 2019). As Manjana Milkoreit states, "our imagination is to a large extent bound to the systems we live in" (2016, 172). Climate fiction, she believes, may be just the tool to strengthen humans' imaginations (Milkoreit 2016, 172). If this is the case, then Australian climate fiction and, by extension, Australian ecofiction is well placed to improve readers' understandings of the impacts of climate change not just on human communities, but on those of the nonhuman as well.

This being said, I do not believe that an explicit focus on climate change is characteristic of all Australian ecofiction; that is, while all Australian climate fiction is Australian ecofiction, not all Australian ecofiction is climate fiction. Novels such as Wright's *The Swan Book* (discussed in Chap. 3), McKinnon's *Storyland* and Mireille Juchau's *The World Without Us* (2015) (discussed in Chap. 4) are frequently cited as examples of the climate fiction, or cli-fi, trend seen in both the Australian publishing industry and globally over the past decade. Other novels, though, such as Harper's *The Dry* and Chris Hammer's *Scrublands* (2018) (Chap. 6) and Anita Heiss' *Bila Yarrudhanggalangdhuray* (2021) (Chap. 7) are not texts that are explicitly linked to the climate fiction genre. While such texts are environmentally themed, their primary plot is not overtly concerned with the climate crisis. Some might say that this distinction is irrelevant and that any twenty-first-century novel even tangentially concerned with environmental issues is in some way contributing to the climate fiction genre, but it is a distinction worth making in the broader context of postcolonial literary representations. The crime texts discussed in Chap. 6, for example, still offer important perspectives of the Australian nonhuman, particularly in their emphasis on drought and bushfire, and these perspectives still reveal vital insights into the continuing dominance of Australia settler attitudes towards nature and, by extension, climate. Scholarly considerations of these kinds of texts that may not be referred to explicitly as climate fiction are therefore vital.

2.3 Indigenous Storytelling and the Nonhuman

Australian literature has a long tradition of evoking vivid depictions of the continent's native habitats. Non-Indigenous Australian writers from May Gibbs and Banjo Paterson to Tim Winton and Kate Grenville tell stories that are reliant on portrayals of quintessentially Australian environments and landscapes. From rainforest and woodland to golden beaches and sweeping farmland, these are the spaces often considered the country's most vibrant and habitable areas. In colonial and postcolonial Australian literature, portrayals of Australian habitats are often limited to the aesthetically pleasing and historically pertinent areas of a country that, in reality, possesses a much wider range of habitats.

Indigenous Australian author Claire G. Coleman's novel *Terra Nullius* (2017) highlights this issue, depicting the allegorical struggle of an alien species that has invaded Earth and is attempting to inhabit a place they are not acclimated to or fond of. In particular, Coleman emphasises the inability of this species to reside in Australia's desert habitats, whilst the humans of Earth (representative of Indigenous Australians) know this place well. This narrative highlights the romanticised view of the Australian natural world present in the nation's early non-Indigenous literature that often obscures the true feelings of Europeans living on the Australian continent. For many non-Indigenous Australians then, and for many still, deserts are wastelands, the deep Australian bush is an inconvenient obstacle to agriculture, and native grasslands are simply paddocks. Roslynn D. Haynes argues that Australian deserts represent a "'before-us' quality"; she observes that Europeans' perspectives of these deserts stem from "the settlers' desire for more pastoral and agricultural land" (1998, 1). The desert, then, often appeared as featureless to early European explorers; it was "a vast and monotonous sameness" (Haynes 1998, 4). For many non-Indigenous Australians today, this stereotype has continued, with the desert and other native Australian habitats often viewed as an ancient, "before-us" place known intimately by some groups of Indigenous Australians but underappreciated by most non-Indigenous people, playing off the perspectives of those early European settlers.

Graham Huggan argues that, even now, Australian literary criticism remains restricted by "its reliance on national(ist) tropes and mythologized binary oppositions" and that various strands of Australian literature are still struggling to move away from their colonial roots (2007, ix). These roots dictate that nature and culture exist in opposition to one another, and that the treatment of the country's native habitats is still dominated by European norms. Australian ecofiction novels of the early twenty-first century increasingly resist the allure of the Romantic tradition when it comes to nature—depicted natural habitats are not always aesthetically pleasing areas of untouched wilderness, and they are not always habitats that are widely familiar.

While it can be argued that Romanticism positioned nature and the nonhuman in ways that were, as Kate Rigby describes, resistant to "narrowly instrumentalist views of the other-than-human" (2008, 5), the movement still largely relied on anthropocentric perspectives of the natural world. Andrew Stables notes that the Romantics typically "mystified nature" (2009, 13); this mystification suggests that, within the Romantic tradition, the nonhuman exudes a certain sacred sublimity that exists for humans to contemplate and take pleasure in. As Onno Oerlemans comments, orthodox Romantics contend that "Nature as it is perceived and represented always signifies; it is an entity or field of entities that must be penetrated and interpreted" (2004, 33). In other words, it simply exists to be understood by the human. The natural places celebrated by the Romantics—like "the mountaintop [and] the ancient forest" (Cronon 1996, 22)—neglected those places that, especially in an Australian context, were deemed desolate and devoid of life by European standards. As will be discussed in this chapter and throughout this book, recent works of Australian ecofiction authored by non-Indigenous writers engage with the Australian nonhuman in a way that does not mystify nature, but rather seeks to understand it as it exists materially, through a more-than-human lens.

For tens of thousands of years, Indigenous Australian storytelling traditions have incorporated the deep connections felt between humans, native habitats, and their other inhabitants. This is a model which could influence non-Indigenous authors to engage more readily with the natural world and the human relationship with it, and to shift the focus away from stories of humans living *on* the natural world to rather living *in* it. Practices of Indigenous Australian oral storytelling, often referred to as yarning, are described by Lynore K. Geia et al. as a dynamic process:

> The stories remain in our conscious state like a thread hanging, waiting to be picked up again, to be continued, reconstructed, reinforced and once again embedded in our ontology. (2013, 15)

The stories passed on through yarning include important aspects of Indigenous history up until and including the present. They also do not exist in a vacuum; Geia et al. describe how "yarns are rarely an individual construct; they carry within them the shared lived experience of their families, and communities" (2013, 15). They also state that yarning is a way for many Indigenous Australian people to "make sense of [their] lived experience" (2013, 15). Ecofiction generally carries the capacity to aid readers' understandings of their and others' experiences in an ecocatastrophic world, and the model of Indigenous storytelling may play a role in the further development of this genre.

Salhia Ben-Messahel argues that the stories of Australia's Indigenous peoples have been largely excluded from the continent's literary canon (2011, 78). Oral Indigenous stories have, in the past, not been properly recognised as being of merit within Australian literary discourse. There are a variety of reasons for this, one of which is that oral stories have been seen through a colonial lens as "a sign of the backward, primitive and ignorant", as Penny van Toorn describes (2000, 20). Additionally, the translation and dissemination of these stories to those outside of Indigenous communities is by no means a straightforward process, and for good reason. Van Toorn explains that while certain communication technologies provide an opportunity for these stories to be more widely read by the Australian populace, they "can also act as a force for changing the content of oral narratives and disrupting the customs that traditionally regulated direct oral transmission" (2000, 20).

While these issues were noted in the early 2000s and the value of Indigenous oral stories within Australian literature has since been more consciously recognised (the website of Indigenous Australian publisher Magabala Books, for example, highlights the importance of collecting and publishing Indigenous oral narratives through their cultural projects for the benefit of future generations (2025)), the exclusion of orality and Indigenous works generally from the Australian canon is still a significant issue today, as are the presumptions exhibited by settler-colonial publishers and readers regarding such works. For example, in 2015 Palyku author Ambelin Kwaymullina lamented the disadvantages experienced by Indigenous authors in Australia today; how they must contend with non-Indigenous editors and reviewers "judging their book against the only other book by a writer of that background that they've ever read". This is further problematised by the fact that, as Kwaymullina states, "for Indigenous peoples, ... the narratives told *about* us vastly outnumber those told *by* us" (2015; italics in original). Additionally, Indigenous knowledge and stories that stem from oral tradition are protected by fraught intellectual property laws that are "based in a [settler-colonial] culture that is individualistic, results-oriented and steeped in written tradition" (Kwaymullina 2015). Based on Kwaymullina's comments, it is clear that Indigenous storytelling practices are still largely undervalued and misunderstood by many non-Indigenous Australians.

With this in mind, what of Indigenous Australian writers of ecofiction, like Wright and Heiss, who are engaging with the nonhuman in a culturally intimate way? In comparison to non-Indigenous Australian writers, how do Indigenous writers depict the nonhuman and what are the potential meanings of such depictions? Mudrooroo[6] contends that for Indigenous writers, the simple act of writing and deciding on the style of said writing—a style that may be vastly different to the norm sought by Australian publishing houses—is inherently political. Often, he says, the Indigenous Australian writer must adapt their style and content to meet "the standards of publishers who have their eyes on the market-place" (1997, 41). Whilst some established Indigenous authors such as Wright may not have this problem now (in many ways, *The Swan Book* steps outside the boundaries of postmodern literary fiction), that is not to say that the issue is not still pervasive, especially for emerging Indigenous writers. Mudrooroo argues that part of the challenge for Indigenous writers approaching mainstream publishers with works written in an atypical style is that

> the Anglo-Celtic readership find it hard to work through the layers of non-compromising Indigenality with the same dedication they bring to European writers such as James Joyce. (1997, 41)

In the realm of ecofiction, then, how does this apply? We see authors like Wright reinventing what we know as Australian literature, making space for more Indigenous Australian writers to decolonise both literature and contemporary environmental discourse in a bid to enact real change. However, ecofiction is also a space in which both Indigenous and non-Indigenous Australian authors can more deeply explore what it means to live in the Anthropocene, pushing aside Romantic conventions of Australia's natural world and making room for a literature which seeks to alter the way readers view their relationship with the Australian nonhuman.

The strong influence of Indigenous Australian thinking within schools of ecocriticism and the environmental humanities should not be understated, in particular the concept of Country which informs some of my discussion about the nonhuman in this book. This concept emphasises the importance of custodianship of and connection to land. Murri Elder Aunty Joan Tranter explains that

[6] As Ben Etherington (2020) describes, there has been past controversy regarding Mudrooroo's claims to Indigeneity. This controversy has never been settled, and so it should be noted that Mudrooroo's comments on Indigenous writing and literature may be considered by some to not be representative of Indigenous perspectives.

...we are referring to land known as *Country* with which each family or language group has held traditional custodianship and responsibilities that have been passed down through hundreds of generations over many, many thousands of years.

We have held these responsibilities to our lands to ensure their spiritual, cultural and economic survival since time began. The land is the source of our identity. Although nowadays many of us may no longer live on our traditional lands and Country, we are still very much connected to it. (2013, 11)

This connection to land extends to a relationship with the nonhuman inhabitants of it. Authored by the Gay'wu Group of Women (five Yolnu women from North East Arnhem Land and three non-Indigenous women who have collaborated on cultural projects for over a decade), *Song Spirals* describes these myriad connections in the following way:

Country is home, it sings to us and nourishes us. It is the feeling of home, the feeling of the seasons that communicate with us. It is all the beings of home. It is everything that we can touch or feel or hear or sense, and it is everything beyond that too. It is everything that belongs in Country, with Country and as Country, including us. And it is the relationships between all those beings too. We come into being together. (2019, 23)

Significantly, multispecies scholars within the environmental humanities are informed by Indigenous scholarship on the natural world. Métis scholar Zoe Todd argues that "Indigenous peoples played and play an integral role in bringing the topic of climate change to the international stage" (2016, 5). European scholarship on the nonhuman is deeply indebted to Indigenous understandings and scholarship of the same. Todd describes that climate, in particular, has long been viewed by Indigenous thinkers as "*a common organizing force*" (2016, 8; italics in original), whilst it is only more recently that European scholars have begun to engage with the idea of climate and the more-than-human as agential powers in their own right. The latter scholars have both consciously and unconsciously laid claim to such thinking, even though these understandings of nature and the nonhuman have run deep within various Indigenous cultures for millennia (Todd 2016).

Thom van Dooren and Deborah Bird Rose's development of "lively ethographies" (2016) speaks to this deep history of Indigenous thought behind climate change and the nonhuman. These ethographies engage with conventional ethnographic theory in an attempt to understand the nonhuman through Indigenous relationships with local species and places. Van Dooren and Rose provide some examples of these, such as understanding what "the disappearance of flying foxes mean[s] to *this* community of Aboriginal people in Australia for whom the species is kin" (2016, 86; italics in original). Further to this, Thom van Dooren et al. are careful to highlight that the term "*species*" within "multispecies studies" should not necessarily be understood as referring to the taxonomy of Western science (2016, 5; italics in original). This suggests that the meaning of the term can be extended to Indigenous understandings of and relationships with animals, plants and other nonhuman beings.

Relationality amongst humans, and between humans and nonhumans, is, of course, complex, and multispecies scholars respond to this; van Dooren et al. explain that "while humans may all be connected to others, they are not all tangled up in the same ways" (2016, 14). Many human cultures engage with the nonhuman, but it cannot be said that all of these cultures are uniformly entwined with the nonhuman. Non-Indigenous Australians, for example, may claim a deep connection to the land they have grown up on and the native nonhumans they have lived amongst, but such a connection cannot be equated to the vastly different relationship between Indigenous Australians and Country. In order to encompass a range of perspectives on human–nonhuman relations as they are represented in the selected Australian ecofiction texts, I draw on Western reflections on the nonhuman, such as the work of Rose and van Dooren, while at other times I turn to Indigenous thought and scholarship, such as the work of Todd and Indigenous Australian scholars Jeanine Leane and Evelyn Araluen.

Val Plumwood's work on philosophical animism will be a key component of my analysis of the nature/culture and human/nonhuman dualisms present in the selected texts. Rose describes how more anthropologists are concerned with rethinking the idea of animism in a way "that engages with Indigenous people's own account of personhood" (2013b, 98). Plumwood, she explains, "was committed to challenging the narrative of mind/matter dualism which allocates all mind to humans, and leaves all the rest of the living and non-living world in a state of mindless matter" (2013b, 95). Entrenched in ideas of philosophical animism are the cultural signifiers used by Indigenous groups around the world to refer to and understand their land and the nonhumans living on it. As Rose points out, a well-known Indigenous Australian example of this is the aforementioned concept of "Country": a word which does not simply describe the physical environment, but a "local multispecies belonging" and a place where there is no divide between nature and culture (2013b, 100). In her work on "Anthropocene noir", Rose also highlights the links between the destruction of Country and the governmental control of Indigenous peoples by colonisers, and the shift away from reconciliation between non-Indigenous and Indigenous Australians in the age of the Anthropocene. She states that "reconciliation is not solely an inter-human project, and relations between Indigenous people and settler-descended people are cross-cut by the dynamics of wider earth systems" (2013a, 208): a concept that I believe Wright engages with strongly in *The Swan Book*.

The Swan Book also considers the concept of "making kin" with nonhuman species: an idea that has been explored extensively by Donna Haraway. Specifically, Haraway's work in cyborg theory and companion species theory has largely influenced how others regard "the human and nonhuman … nature and culture" (2003, 4), with some of her more recent writing falling within multispecies studies. In her paper "Anthropocene, Capitalocene, Plantationocene, Chthulucene", Haraway grapples with the complexities of "kin-making" and of "kinds" rather than taxonomied "species" (2015a, 162). Haraway argues that her "purpose is to make 'kin' mean something other/more than entities tied by ancestry or genealogy" (2015a, 161). This concept highlights that it is not just species similar to us biologically or historically that deserve these connections; these special relationships we form with "kin"—anything else that we share this planet with—represent something closer to an emotional, ethical choice rather than an evolutionary one.

Haraway, too, recognises the significance of Indigenous knowledge in her quest to more closely align the human with the more-than-human. She states that "decolonial indigenous peoples and projects are central to my stories of alliance" (2016, 71). These "stories of alliance", as Haraway describes, represent the coming together of the sciences and the arts to create ways of understanding the multispecies worldings which are essential to kin-making (2016, 71). These science-art projects do not just involve Euro-Western thinkers but also, and more importantly, the Indigenous peoples who have long engaged with the natural places and nonhuman beings relevant to such projects, and who can provide deep understandings of what it means to make kin with nonhumans.

2.4 Multispecies Studies and Nonhuman Symbolism

As will already be clear, "nonhuman" is a term that I use in a variety of ways in my analysis; I consider nonhumans as individuals, as species, as habitats and as phenomena. Although I acknowledge that nonhuman experience exists in meaningful ways outside of the perceptions of humans, it is, of course, true that nonhumans play significant roles in both Indigenous and non-Indigenous Australian cultures. The relationships formed between humans and nonhumans, especially in the context of the climate crisis, provide opportunities for Australians to more deeply appreciate the human impact on their natural surroundings and what this means for their own and other living beings' communities.

Whilst the native nonhuman—those species and habitats endemic to the Australian continent—may be one focus of my discussions of nonhuman representation in Australian ecofiction, it is by no means the only focus. Non-native nonhumans form another facet of my analysis, with a particular emphasis on those nonhumans that play a central role in Australia's Anglo-agricultural practices. Additionally, I also utilise "nonhuman" to describe certain nonliving, or abiotic, nonhumans such as drought, fire and flooding, which I refer to more broadly as "phenomena". Significantly, the idea that these phenomena are completely natural is now largely outdated, with evidence asserting that anthropogenic climate change plays a role in causing intense drought and bushfire (Jones et al. 2013, 2). However, these events still exist externally to human societies and cultures and their nonhuman-ness forms an important part of my close-reading analysis, especially in my discussion of Australian crime fiction in Chap. 6.

Returning briefly to the native nonhuman, this may also refer to the broader idea of habitat, rather than simply individual Australian species. Throughout this book, I often use the term "natural world" to refer to the wider natural environment or "nature" (as discussed earlier), and the term "native habitat" or "habitat" when referring to the natural places endemic to Australia, such as grassland, rainforest and marine habitats. These habitats are usually inhabited by the nonhuman, visited by the human, or both simultaneously, and so the term "habitat" will also refer to those natural places which have been altered by the human whilst still retaining their sense of "the natural" (for example, the surrounding rainforest of farming landscapes, depicted in Mireille Juchau's *The World Without Us* (2015)).

It is therefore necessary to indicate how this term "habitat" may relate to the more human-centric idea of "landscape". This latter term is an important theme in all of the texts discussed in this book, but needs to be defined in a way that reveals both its separation from the nonhuman, and its connection to it. In particular, landscape plays a critical role in my analysis of the crime fiction texts of Chap. 6. The landscapes depicted in these texts are often rural and agricultural in nature, such as the landscapes of farming towns and the surrounding countryside, and the often drought-stricken, dry landscapes of the farms themselves. These differences between habitat and landscape can be linked to Potter's discussion of an imaginary of the ground. She speaks of "an obsession with ground" within settler-colonial Australia; in particular, "a ground's given qualities of flatness, sharpness, unevenness, and instability (in essence, its amenability to domestication)" (2019, 5). In this sense, one can consider the ground of what we know now as Anglo-agricultural landscapes to be the transformation, or domestication, of the ground of native Australian habitats.

Multispecies studies is a rapidly growing field that can assist with these enquiries about nonhuman representation, with the work of Australian scholars such as van Dooren and Rose often at the forefront of such enquiries. Multispecies studies is interdisciplinary in nature, bringing together various disciplines in an attempt to better understand the human and the nonhuman. Van Dooren et al. describe this aptly:

Multispecies studies takes up this understanding of our world, drawing inspiration from the natural sciences and beyond, bringing diverse bodies of knowledge into conversation and pushing them in new directions. Multispecies scholars are asking how human lives, life ways, and accountabilities are folded into these entanglements. (2016, 2)

They also comment on how multispecies studies goes beyond the idea of the nonhuman as simply symbolic of human experience, instead claiming that "scholars in multispecies studies have aimed to provide 'thick' accounts of the distinctive experiential worlds, modes of being, and biocultural attachments of other species" (2016, 6). Whilst a productive endeavour, I would argue that it is by no means a simple one. In the discussion of the exemplar novels across Chaps. 3–7 of this book, it has been difficult to exclusively consider the lives of the nonhuman without often foregrounding the lives of humans in the process. Part of my research involves an exploration of how Australian ecofiction authors use nonhumans as representational devices symbolising human experiences of ecological crises (and how this is often problematic), and so it is not always in my interests as a researcher to completely disregard the symbolic nature of nonhuman representation in my chosen texts. I recognise this symbolism as problematic, and it is an issue which is further discussed in Chap. 3, but a clarification is also warranted here. Nonhuman symbolism is a practice that has been problematised by many working in critical animal studies over the years. Josephine Donovan is one scholar who comments on the problem of analysing animal representation in literature from ideological standpoints:

...literary critics have begun reexamining the status of animals in literature, identifying ideologically driven representation, questioning the ubiquitous aesthetic exploitation of animal pain and suffering... (2016, 95)

With this in mind, my close-reading analysis attempts to move past such projections of human experiences onto the nonhuman, and instead provide a nuanced understanding of how nonhumans are represented in these texts, both as problematically symbolic and as actual characters with stories in their own right. Multispecies studies informs these considerations, as I explore how the reader can better understand the human impact on the natural world by comprehending the very real nonhuman experiences of ecological catastrophe.

In her work defining the Chthulucene, Haraway also contributes to these discussions, writing that "only with intense commitment and collaborative work and play with other terrains, flourishing for rich multispecies assemblages that include people will be possible" (2015a, 160).[7] This concept is explored by some of the authors discussed in this book, but if such "collaborative work" is to be achieved with these "other terrans" (Haraway 2015a, 160), then what of the issue of agency? Haraway's implication is that if humans are to work more collaboratively with the more-than-human, then a better understanding of nonhuman agency is required. In *The Companion Species Manifesto* (2003), Haraway continuously

[7] It is important to mention here some of the critique of Haraway's work, particularly in terms of her claims around the importance of collaborating with other species, such as those in the agricultural sector. Melissa Boyde suggests that Haraway's approach fails to interrogate the sacrifice of some nonhumans who "collaborate" with humans, emphasising the pain and violence experienced by many in farming industries, particularly dairy cows whose calves are often forcibly separated from them. Boyde says of Haraway's reflections that "naming the practices of the dairy industry to be the cows and humans 'working together' ... with the breeder seems a stretch of the imagination" (Haraway 2015b, 7, cited in Boyde 2018, 21–22). Rather, Boyde argues, it may be more accurate to describe the experiences of nonhumans trapped in these industries as living through loss and "practicing the art of war" (2018, 22).

returns to the importance of defining agency. Exploring the significance of human–dog relationships, Haraway remarks that

> [i]n both popular and professional dog worlds what is at stake is two-fold: 1) the relation between what counts as nature and what counts as culture in Western discourse and its cousins, and 2) the correlated issue of who and what counts as an actor. (27)

As aforementioned, Australian fiction is undoubtedly influenced by the nation's colonial roots and its still largely colonial present, and a problematic nature–culture dichotomy is very much a part of most non-Indigenous, Anglo-Celtic Australian literature. I draw on existing scholarly discussions within the environmental humanities that are concerned with this dichotomy to critique this barrier and its representation in Australian ecofiction authored by non-Indigenous writers. The issue of agency described by Haraway ("who and what counts as an actor" (2003, 27)) is also a common theme in Australian ecofiction texts, with many authors exploring what it means to be human in a world where other beings possess an agency that is vastly different to the agency that settler-colonisers know.

Val Plumwood's idea of "philosophical animism" comes into play here. Rose describes how Plumwood was concerned with "understanding that the living world is powerful and possesses its own agency and sentience" (2013b, 94). Such an idea can of course be largely credited to various Indigenous cultures around the world prior to Plumwood's exploration of it, which is discussed in detail by Rose following interviews she conducted with Indigenous Australians. Rose highlights this further in her work on the concept of "Anthropocene noir" in which, as previously quoted, she describes the act of reconciliation[8] as "not solely an inter-human project" (2013a, 208).

Globally, Indigenous thinkers have come to similar conclusions long before non-Indigenous ecocritics, environmental philosophers and environmental humanists ever did. Particularly regarding climate change, many Indigenous thinkers acknowledge that environmental and cultural destruction has been experienced by many Indigenous societies worldwide as a result of colonisation. Potawatomi scholar Kyle Powys Whyte contends that

[8] Catriona Elder describes reconciliation as "the key state intervention" proposed by Australia's 1991 Federal Parliament to address the range of "legal and political issues" surrounding "the need for proper recognition of Indigenous peoples by the state" (2017, 74).

> [d]ifferent forms of colonialism…, whether through environmental destruction, land dispossession or forced relocation, have ended Indigenous peoples' local relationships to thousands of plants, animals, insects, and entire ecosystems. (2018, 226)

While many non-Indigenous Australians continue to develop more distant relationships with the natural world in their everyday lives (whether intentionally or not), it must be noted that Indigenous Australian peoples have been forcibly disconnected from nature and the nonhuman by colonisers over the past three centuries. Whyte goes on to state that "some Indigenous peoples … offer the idea that we confront climate change having already passed through environmental and climate crises arising from the impacts of colonialism" (2018, 226).

It is perhaps no surprise, then, that many scholars advance the notion that Indigenous epistemologies should play a more central role in multispecies understandings within the environmental humanities (as well as the ecological sciences), as these knowledges encompass the deeply felt experiences of peoples who have already endured environmental apocalypse. Academics are therefore asking what can be learnt from such epistemologies in the context of the climate emergency. Todd asserts that "Indigenous peoples … play an integral role in bringing the topic of climate change to the international stage" (2016, 5). Similarly, and based on the above contention from Whyte, Indigenous peoples are well placed to address the increasing disconnect between the human and nonhuman: they too have experienced this disconnection, albeit forcibly, but have still retained culturally important relationships with the land and water, and other living inhabitants (2018, 226).

Tranter reflects on this strength of Indigenous connection to lands and waterways in spite of the disconnection enforced by colonisers. She states that "although nowadays many of us may no longer live on our traditional lands and Country, we are still very much connected to it" (2013, 11). This perspective is particularly important in the context of Wright's *The Swan Book,* which is discussed in Chap. 3. In *Dark Emu* (2014), Yuin, Bunurong and Tasmanian man Bruce Pascoe describes how the law of Indigenous Australian cultures insists that

> …the land was held in common and that people were the mere temporal custodians. Individuals were responsible for particular trees, rivers, lakes, and stretches of land, but only so these could be delivered forward to the next generation in accordance with law. (198–199)

Pascoe goes on to describe how such a system of custodianship is viewed as a kind of "jigsaw mutualism": a means of caring and connecting with "particular pieces of the jigsaw … so that it added to rather than detracted from the pieces of their neighbours and the epic integrity of the land" (2014, 199). This represents a vastly different perspective of the Australian natural world and its nonhuman inhabitants to that of European colonisers who divided the land for individual, material gain. My close-readings of the selected texts therefore encompass theory relating to multispecies studies (amongst other theory from within the environmental humanities), and Indigenous epistemologies regarding human–nonhuman relationships—in particular, Indigenous Australian cultural understandings of Country.

In her discussion of Plumwood's philosophical animism, Rose discusses whether nonhumans may "experience an emotional life … have a sense of morality and justice … [or] experience empathy" (2013b, 97). She summarises Plumwood's exploration of animism as a means of understanding "attentive interactions in the sentient world" (2013b, 93), suggesting that the world outside of the human also experiences sentience. She also describes how Plumwood asserts "that as other creatures live their lives, so they communicate aspects of themselves" (2013b, 98). Significantly, Rose draws on Indigenous epistemologies in her analysis of Plumwood's philosophical animism, based on interviews with Indigenous Elders. In this work, one Elder describes how "birds got ceremony of their own—brolga, turkey, crow, hawk, white and black cockatoo—all got ceremony, women's side, men's side, … everything" (2013b, 100).

This indicates an absence of the nature/culture divide within Indigenous epistemologies (Rose 2013b, 100)—something that is otherwise commonly observed in Western modes of thinking regarding the natural world. The idea that nonhumans experience their own ceremonies suggests that they too may communicate their own stories. In Plumwood's work on shadow places, she suggests that constructing "a more unified place relationship" (2008, 143)—something that is common in Indigenous cultures—is a means by which we can increase sensitivity towards both the nonhuman and other humans. Such a relationship is more ecologically aware in its confrontation of Western nature–culture dualisms, such as "mind/body, reason/emotion and respect/use" (Plumwood 2008, 143). The Gay'wu Group of Women reflect a sense of nonhuman agency in their thoughts on "spirit beings" in Yolnu culture:

The animals, the spirit beings, the stars and space, everything has its knowledge and Law. The morning star has its own understanding, and the evening star has its own understanding.... We have to stop and think what we are doing and try to listen. The spirit will tell us, the land will tell us. (Burarrwana et al. 2019, 153–154)

Such an understanding engages with a unified relationship between human and nonhuman rather than a dualistic perspective of humans and the land.

It is these kinds of relational ontologies surrounding the more-than-human that are key to my conceptualisation of the postcolonial eco-novel in this book. Not all of the selected texts necessarily adhere to material, relational depictions of the more-than-human; in fact, some actively work against this notion, instead exploiting the more-than-human as "a blank screen for the projection of human meaning" (Armstrong 2008, 3, cited in Donovan 2016, 95). It is key, however, to understand what such literary exploitation means in the broader context of Western (in)action on global ecological crisis—and what positive cultural and environmental transformations are possible within and through those novels that may be more relational in their approach to the more-than-human, in turn challenging dominant colonial ideologies of nature in postcolonial nation states like Australia.

References

Armstrong, Philip. 2008. *What Animals Mean in the Fiction of Modernity*. Milton Park: Routledge.

Bal, Reyyan. 2021. Nature vs. "Reality" in Fantasy Fiction: The Potential for Ecocritical Imaginings. *Gaziantep University Journal of Social Sciences* 20 (3): 1283–1296.

Barrera, Mario. 1979. *Race and Class in the South-West: A Theory of Racial Inequality*. Notre Dame: University of Notre Dame Press.

Ben-Messahel, Salhia. 2011. Reconfiguring Australia's Literary Canon: Antipodean Cultural Tectonics. *Commonwealth: Essays and Studies* 34 (1): 77–91.

Biermann, Soenke. 2011. Knowledge, Power and Decolonization: Implication for Non-Indigenous Scholars, Researchers and Educators. *Counterpoints* 379: 386–398.

Boyde, Melissa. 2018. The Dairy Issue: "Practicing the Art of War". *Animal Studies Journal* 7 (2): 9–24.

Bradley, James. 2015. *Clade*. Melbourne: Hamish Hamilton.

Burarrwana, Laklak, Ritjilili Ganambarr, Merrkiyawuy Ganambarr-Stubbs, Banbapuy Ganambarr, Djawundil Maymuru, Sarah Wright, Sandie Suchet-Pearson, and Kate Lloyd. 2019. *Song Spirals: Sharing Women's Wisdom of Country through Songlines*. Crows Nest: Allen & Unwin.

Clark, Timothy. 2014. Nature, Post Nature. In *The Cambridge Companion to Literature and the Environment*, ed. Louise Westling, 75–89. New York: Cambridge University Press.

Coleman, Claire G. 2017. *Terra Nullius*. Sydney: Hachette Australia.

Cooke, Stuart, and Peter Denney. 2021. Thinking About Transcultural Ecocriticism: Space, Scale and Translation. In *Transcultural Ecocriticism: Global, Romantic and Decolonial Perspectives*, ed. Stuart Cooke and Peter Denney, 13–31. London: Bloomsbury Academic.

Cronon, William. 1996. The Trouble with Wilderness: Or, Getting Back to the Wrong Nature. *Environmental History* 1 (1): 7–28.

Donovan, Josephine. 2016. *The Aesthetics of Care: On the Literary Treatment of Animals*. New York: Bloomsbury.

Doyle, Briohny. 2016. *The Island Will Sink*. Melbourne: The Lifted Brow.

Elder, Catriona. 2017. Unfinished Business in (Post)Reconciliation Australia. *Australian Humanities Review* 61: 74–93.

Etherington, Ben. 2020. The Living and the Undead. *Sydney Review of Books*, April 29.

Forter, Greg. 2023. Towards the Cli-Fi Historical Novel; Or, Climate Futures Past in Recent Fiction. *American Literary History* 35 (4): 1665–1687.

Forthomme, Claude. 2014. A Chat with the Man Who Dreamt up Cli Fi. *Impakter*, July 30. https://impakter.com/climate-fiction-a-chat-with-the-man-who-coined-the-term-cli-fi/. Accessed 8 Feb 2025.

Geia, Lynore K., Barbara Hayes, and Kim Usher. 2013. Yarning/Aboriginal Storytelling: Towards an Understanding of an Indigenous Perspective and Its Implications for Research Practice. *Contemporary Nurse: A Journal for the Australian Nursing Profession* 46 (1): 13–17.

Gelder, Ken. 2001. Recovering Australian Popular Fiction: Towards the End of Australian Literature. In *Australian Literary Studies in the 21st Century*, ed. Philip Mead, 112–120. Association for the Study of Australian Literature.

Gelder, Ken, and Rachael Weaver. 2008. Colonial Australian Crime Fiction. In *The Anthology of Colonial Australian Crime Fiction*, ed. Ken Gelder and Rachael Weaver, 1–10. Carlton: Melbourne University Press.

Ghosh, Amitav. 2016. *The Great Derangement: Climate Change and the Unthinkable*. Chicago and London: University of Chicago Press.

Ghosh, Amitav. 2019. *Gun Island*. New York: Farrar, Straus and Giroux.

Grant, Stan. 2016. *Talking to My Country*. Sydney: HarperCollins Publishers.

Haraway, Donna. 2015a. Anthropocene, Capitalocene, Plantationocene, Chthulucene: Making Kin. *Environmental Humanities* 6 (1): 159–165.

Haraway, Donna. 2015b. Preface. *Angelaki: Journal of the Theoretical Humanities* 20 (2): 5–14.
Haraway, Donna. 2016. *Staying with the Trouble: Making Kin in the Chthulucene*. Durham: Duke University Press.
Haraway, Donna. 2003. *The Companion Species Manifesto: Dogs, People and Significant Otherness*. Chicago: Prickly Paradigm Press.
Hammer, Chris. 2018. Scrublands. Crows Nest: Allen & Unwin.
Harper, Jane. 2016. *The Dry*. Sydney: Pan Macmillan Australia.
Haynes, Roslynn D. 1998. *Seeking the Centre: The Australian Desert in Literature, Art and Film*. Cambridge: Cambridge University Press.
Heiss, Anita. 2021. *Bila Yarrudhanggalangdhuray River of Dreams*. Sydney: Simon & Schuster.
Huggan, Graham. 2007. *Australian Literature: Postcolonialism, Racism, Transnationalism*. Oxford and New York: Oxford University Press.
Hughes-D'Aeth, Tony. 2023. Australian Fiction in the Anthropocene. In *The Cambridge Companion to the Australian Novel*, ed. Nicholas Birns and Louis Klee, 289–304. Cambridge: Cambridge University Press.
Jemisin, N. K. 2015. *The Fifth Season*. London: Orbit.
Jemisin, N. K. 2016. *The Obelisk Gate*. London: Orbit.
Jemisin, N. K. 2017. *The Stone Sky*. London: Orbit.
Jones, Roger, Celeste Young, John Handmer, Adrian Keating, Gayathri Mekala, and Peter Sheehan. 2013. *Valuing Adaptation Under Rapid Change*. Gold Coast, QLD: National Climate Change Adaptation Research Facility.
Juchau, Mireille. 2015. *The World Without Us*. London: Bloomsbury Publishing.
Kwaymullina, Ambelin. 2015. Let the Stories In: On Power, Privilege and Being an Indigenous Writer. *The Wheeler Centre: Notes*, June 12. https://www.wheelercentre.com/news-stories/2015/let-the-stories-in-on-power-privilege-and-being-an-indigenous-writer. Accessed 28 Jan 2025.
Leane, Jeanine. 2010. Aboriginal Representation: Conflict or Dialogue in the Academy. *The Australian Journal of Indigenous Education* 39 (Supplement): 32–39.
LeMenager, Stephanie. 2017. Climate Change and the Struggle for Genre. In *Anthropocene Reading: Literary History in Geologic Times*, ed. Tobias Menely and Jesse Oak Taylor, 220–238. University Park: Pennsylvania State University Press.
Magabala Books. *Cultural Projects*. https://www.magabala.com/pages/cultural-projects. Accessed 11 Feb 2025.
Malecki, Wojciech, Piotr Sorokowski, Boguslaw Pawlowski, and Marcin Cieński. 2019. *Human Minds and Animal Stories: How Narratives Make Us Care About Other Species*. New York and London: Routledge.
McKinnon, Catherine. 2017. *Storyland*. Sydney: HarperCollins Publishers.

Milkoreit, Manjana. 2016. The Promise of Climate Fiction: Imagination, Storytelling and the Politics of the Future. In *Reimagining Climate Change*, ed. Paul Wapner and Hilal Elver, 171–191. London: Routledge.

Milner, Andrew. 2020. Eutopia, Dystopia and Climate Change. In *Ethical Futures and Global Science Fiction*, ed. Zachary Kendal, Aisling Smith, Giulia Champion, and Andrew Milner, 77–97. Cham: Palgrave Macmillan.

Milner, Andrew, and J. R. Burgmann. 2018. A Short Pre-History of Climate Fiction. *Extrapolation* 59 (1): 1–23.

Mitchell, David. 2014. *The Bone Clocks*. London: Sceptre.

Modjeska, Drusilla. 2002. *Timepieces*. Sydney: Picador.

Morgan, Sally. 1987. *My Place*. Fremantle: Fremantle Arts Centre Press.

Mrozewicz, Anna Estera. 2020. The Landscapes of Eco-Noir: Reimagining Norwegian Eco-Exceptionalism in Occupied. *Nordicom Review* 41 (Special Issue 1): 85–105.

Mudrooroo. 1997. *Milli Milli Wangka: The Indigenous Literature of Australia*. South Melbourne: Hyland House Publishing.

Oerlemans, Onno. 2004. *Romanticism and the Materiality of Nature*. Toronto: University of Toronto Press.

O'Mahony, Lauren. 2017. Death and the Australian Rural Romance Novel. *Text* 21 (Special 45): 1–14.

O'Mahony, Lauren. 2020. Australian Romance Fiction. In *The Routledge Research Companion to Popular Romance Fiction*, ed. Jayashree Kamblé, Eric Murphy Selinger, and Hsu-Ming Teo, 72–96. London and New York: Routledge.

Pascoe, Bruce. 2014. *Dark Emu: Aboriginal Australia and the Birth of Agriculture*. Broome: Magabala Books.

Plumwood, Val. 2008. Shadow Places and the Politics of Dwelling. *Australian Humanities Review* 44: 139–150.

Potter, Emily. 2005. Ecological Crisis and Australian Literary Representation. *Australian Humanities Review* 37: np.

Potter, Emily. 2015. Programme: Dreaming, Timekeeping and Becoming. In *Assembling the Centre: Architecture for Indigenous Cultures*, ed. Janet McGaw and Anoma Pieris, 100–120. London and New York: Routledge.

Potter, Emily. 2019. *Writing Belonging at the Millennium: Notes from the Field on Settler-Colonial Place*. Bristol: Intellect.

Rigby, Kate. 2008. "Lifting Our Eyes from the Page": Religion, Literature and the Earth: Editorial Introduction. *Religion & Literature* 40 (1): 1–9.

Rodoreda, Geoff. 2023. Mabo, History, Sovereignty: The Contemporary Postcolonial Novel. In *The Cambridge History of the Australian Novel*, ed. David Carter, 421–436. Cambridge: Cambridge University Press.

Rose, Deborah Bird. 2013a. Anthropocene Noir. *Arena Journal* 41: 206–219.

Rose, Deborah Bird. 2013b. Val Plumwood's Philosophical Animism: Attentive Interactions in the Sentient World. *Environmental Humanities* 3 (1): 93–109.

Rugg, Linda Haverty. 2017. Displacing Crimes against Nature: Scandinavian Ecocrime Fiction. *Scandinavian Studies* 89 (4): 597–615.
Schneider-Mayerson, Matthew. 2018. The Influence of Climate Fiction: An Empirical Survey of Readers. *Environmental Humanities* 10 (2): 473–500.
Schneider-Mayerson, Matthew, Alexa Weik von Mossner, W. P. Malecki, and Frank Hakemulder. 2023. *Empirical Ecocriticism: Environmental Narratives for Social Change*. Minneapolis: University of Minnesota Press.
Seran, Justine. 2015. Australian Aboriginal Memoir and Memory: A Stolen Generations Trauma Narrative. *Humanities* 4 (4): 661–675.
Stables, Andrew. 2009. The Unnatural Nature of Nature and Nurture: Questioning the Romantic Heritage. *Studies in Philosophy and Education* 28: 3–14.
Todd, Zoe. 2016. An Indigenous Feminist's Take On The Ontological Turn: "Ontology" Is Just Another Word For Colonialism. *Journal of Historical Sociology* 29 (1): 4–22.
Tranter, Aunty Joan. 2013. Indigenous Country Explained. *Education* 94 (7): 11.
VanderMeer, Jeff. 2014. *Southern Reach*. New York: Farrar, Straus and Giroux.
Van Dooren, Thom, and Deborah Bird Rose. 2016. Lively Ethography: Storying Animist Worlds. *Environmental Humanities* 8 (1): 77–94.
Dooren, Van, Eben Kirksey Thom, and Ursula Münster. 2016. Multispecies Studies: Cultivating Arts of Attentiveness. *Environmental Humanities* 8 (1): 1–23.
Walton, Jo Lindsay, and Samantha Walton. 2018. Introduction to Green Letters: Crime Fiction and Ecology. *Green Letters* 22 (1): 2–6.
Watson, Irene. 2009. Sovereign Spaces, Caring for Country, and the Homeless Position of Aboriginal Peoples. *South Atlantic Quarterly* 108 (1): 27–51.
Whyte, Kyle. 2017. Indigenous Climate Change Studies: Indigenizing Futures, Decolonizing the Anthropocene. *English Language Notes* 55 (1–2): 153–162.
Whyte, Kyle. 2018. Indigenous Science (Fiction) for the Anthropocene: Ancestral Dystopias and Fantasies of Climate Change Crises. *Environment and Planning E: Nature and Space* 1 (1–2): 224–242.
Wright, Alexis. 2013. *The Swan Book*. Artarmon: Giramondo Publishing Company.

CHAPTER 3

Border Transgressions and Human–Animal Relations

3.1 Ecocriticism, Animal Studies and Decolonial Ethics

In their work on multispecies ethnography, Laura A. Ogden et al. describe the nonhuman world as a place of:

> multispecies encounters [that] has its own logic and rules of engagement ... encompassing the flow of nutrients and matter, the liveliness of animals, plants, bacteria, and other beings. (2013, 6)

With this in mind, how, then, can a human understand the life of a nonhuman being—both the multispecies worlds existing independently of humans, and the life lived amongst and alongside humans? This chapter contends that literature provides one means by which humans can attempt to comprehend the animal other, allowing readers to cross the border between anthropocentricism and a potentially more engaged, ecocentric perception of what it is to be more-than-human in the Anthropocene.

Alexis Wright's *The Swan Book* (2013) and Merlinda Bobis' *Locust Girl* (2015) are two texts of Australian ecofiction that provide opportunities for the reader to consider a space and time in which the human becomes animal and the animal becomes human. The protagonists of these works can be construed as messengers to the reader, passing on the knowledge gained from their lived encounters with animals in the hopes of

R. Fetherston, *Theorising the Postcolonial Eco-Novel*, Literatures, Cultures, and the Environment,
https://doi.org/10.1007/978-3-032-04466-2_3

understanding their own personal trauma as well as the trauma experienced by the more-than-human themselves. I argue that within both these texts a sense of communicative, interspecies ecocentrism is expressed, a crossing of multispecies borders in a space where interspecies relationships find opportunities to subvert the dominant human-centric paradigms prevalent within modern discourse on the environment. In both *The Swan Book* and *Locust Girl*, I contend that such an ecocentric ethos is supported: the idea that if humans make the ethical choice to engage with the nonhuman more deeply, then a positive shift towards a less mechanistic perspective of the animal other may be conceivable. Significantly, this shift also acknowledges and challenges the totality of colonisation, particularly through the concept of borders, implying that to move toward a more ecocentric narrative about postcolonial Australia means to move past problematic settler-colonial understandings of nature and the nonhuman.

It is difficult, if not impossible, to define Australian ecofiction without reference to both climate change and Indigenous epistemologies. It is therefore necessary to acknowledge that as a non-Indigenous Australian, my interpretation of such epistemologies as well as a text written by an Indigenous Australian author is limited. The Australian publishing industry is seeing more and more novels which explore the theme of climate change as a real and ever advancing threat, and is also seeing an increase in the publication of fiction authored by Aboriginal and Torres Strait Islander people. Texts such as Wright's *The Swan Book*, Nardi Simpson's *Song of the Crocodile* (2020) and Ellen van Neerven's *Heat and Light* (2014) explore the problematic nature of colonial approaches to living in Australian environments and put forward Indigenous ideas regarding the causes and possible solutions to social, environmental and political problems in Australia, many of which stem from colonisation. In this chapter, I therefore engage with some Indigenous epistemologies while approaching the topic of climate change in modern Australian fiction.

The Swan Book and *Locust Girl* both explore ideas of multispeciesism, emphasising the significance of the animal in the protagonists' understanding of the natural and human world. In this chapter, I discuss how these texts break down the Cartesian logic of nature versus culture, human versus animal, and self versus other through a literary engagement with both animals and borders. Timothy Clark describes how many believe that "the ultimate source of humanity's destructive relation to the natural world is *dualism*..." (2014, 77; italics in original). Wright's and Bobis' novels disintegrate these dualisms, both of them challenging these traditional Western models not only through their depiction of environmental devastation but through an engagement with non-Western storytelling practices in the literary form. *The Swan Book* in particular is intimately informed by the complex ontologies and epistemologies of Indigenous Australians which do not support the idea of a nature/culture dichotomy.

Both texts also portray the challenge of shifting towards a more ecocentric and less anthropocentric worldview; to invest more of the Western humanist self into a relationship with the nonhuman requires sacrificing Euro-Western understandings of what it is to be human, and of nature and the nonhuman. So, whilst connecting with the nonhuman other—specifically the native nonhuman other—may be beneficial, it is by no means an easy fix. The protagonists of each novel—Oblivia in *The Swan Book* and Amedea in *Locust Girl*—are effectively punished for their ability to connect with the nonhuman. Their relationships with other species ostracise them from human society, leaving them to fend for themselves in the climate dystopias which each text portrays. It is their unique connection to the nonhuman animal, though, that provides these protagonists with a different means of storying climate change, framing it as a crisis that goes far beyond the human in terms of its impact and possible solution. The influence of local stories within the narratives (that is, the stories told amongst the characters themselves about the places they inhabit) is of particular importance in both *The Swan Book* and *Locust Girl* and will also be discussed in regard to characters' relationships with the nonhuman animal.

Subsequently, I refer to the protagonists of *The Swan Book* and *Locust Girl* as "messengers". Partly inspired by Donna Haraway's concept of kin-making (2015), I contend that these protagonists interpret the lives and experiences of the animals with which they converse and attempt to pass on the knowledge gained through this communication to others, as well as the reader. Oblivia and Amedea are human, but the connections they form with animals—the black swans in *The Swan Book* and a locust in *Locust Girl*—symbolise the more-than-human quality of their personhoods, as well as the personhood of the more-than-human. It is this quality which allows Oblivia and Amedea to effectively understand the perspectives of nonhuman agents for the potential benefit of both humans and nonhumans. Their intimate connection to particular animals facilitates a nuanced exploration of animal agency and an attentiveness to how the nonhuman is impacted by climate change. It is significant that their messages are often garbled, distorted by the inherent difficulty of communicating on behalf of other beings as well as by the competing perspectives on how humans should interact with Australia's natural environments and nonhumans. It is this messiness of the human–animal relationships that undermines the rationalist containment invoked by Western dualisms.

While *The Swan Book* shares Bobis' preoccupation with understanding climate change through the nonhuman animal, Wright's work is strongly informed by Indigenous storytelling traditions and contemporary issues within Indigenous Australian intellectual and political life (Mead 2016, 1). Wright utilises the climate crisis as a means through which to address what Philip Mead describes as a "contemporary socio-political understanding of political reality and narrative, of individual Indigenous activism and nation" (2016, 3). Bobis' work is also concerned with the links between political reality and narrative, although outside of an Aboriginal framework. As a Filipino-Australian author, Bobis has identified herself as "an ethnicised story-maker" and has experienced first-hand the ways in which "Otherness is legitimised, packaged and consumed within the Australian nation", particularly through its national literature (2010, 3). It is not difficult to see some alignment between Bobis' non-fiction critique on this topic and her fiction's approach to borders, containment and movement. As she writes in her essay, "The Asian Conspiracy", which discusses what it means to be an ethnicised writer both within and outside of the Australian settler-colonial nation state's literary boundaries: "How not to be silenced, or how to circumvent the gate: gather a comprehensive view of the landscape. The outsiders have an edge: to survive they must know both outside and inside, both margins and centre" (2010, 5). Her fiction can therefore be read as not just an exploration of geographical border politics, but also the border politics within postcolonial Australia's literary spaces.

Both novels are decolonial in their approach to storying climate crisis through human–animal relationships, and so it is important for me to first situate this analysis within current debates of critical animal studies, particularly regarding the Western-colonial ideologies that have dominated this area of research for some time. Critical animal studies considers the points at which "activism, academia and animal suffering and maltreatment" overlap with one another (Taylor and Twine 2014, 2). The emergence of critical animal studies in the 2000s reflects what Nik Taylor and Richard Twine consider "the start of a long overdue change in contemporary academia that better reflects the importance of other animals as social beings in their own right" (2014, 3). Key to my analysis is the commitment of critical animal studies to position "other animals not as objects or as otherwise instrumentalised entities, but as sentient beings who challenge our notions of respectful interaction at epistemic and ontological levels" (Murray 2022, 69). Jessica Murray asserts that even texts that we may consider to be climate fiction, and therefore are presumably more engaged than other fiction with more-than-human agencies and life-worlds, can still endorse problematic anthropocentric attitudes towards the nonhuman animal (2022, 70)—hence the need for critical animal studies to be employed in more scholarship of climate fiction.

However, Billy-Ray Belcourt has criticised the field of critical animal studies for invisibilising Indigeneity and generally failing "to center an analysis of settler colonialism" in its "anti-speciesist" approach (2020, 20). To Belcourt, incorporating a serious critique of colonialism into critical animal studies is needed if anthropocentrism—a colonial, capitalist and speciesist logic—is to be appropriately challenged (2020, 21). He argues that "neither the abolition of speciesism nor the production of more ethical human-animal relationalities can occur … without the end of settler colonialism" (2020, 20). Multispecies studies, rather than critical animal studies, is, I believe, better positioned to interrogate literary representations of the animal because of its comparatively more explicit attentiveness to more-than-human relationality in decolonial contexts. As Thom van Dooren et al. ask in their overview of this burgeoning field of study, "How do colonialism, capitalism, and their associated unequal power relations play out within a broader web of life?" (2016, 3).

Most of the texts discussed throughout this book are not decolonial in this sense—they may critique parts of the colonial project, but they also emphasise the urgent need for shifts in settler understandings of the nature and the more-than-human. In this respect, the settler is still a settler, living on stolen land, in their attempts to destabilise dominant, exploitative attitudes towards the environment. The novels discussed in this chapter, however, take a decidedly more decolonial approach, challenging not only the colonial project as a whole, but specific notions of property, enclosure, nationhood and anthropocentrism that are at the heart of it.

In my analysis of *The Swan Book*, I refer to postcolonial scholarship on the link between Indigenous genocide and the decimation of Country and its nonhuman inhabitants. James Procter describes postcolonial scholarship as feeling "more at home with reader resistance, reading against the grain, rather than with reading for pleasure or gratification" (2009, 187).

The experience of reading the works of Indigenous Australian writers such as Wright, Claire G. Coleman, Oodgeroo Noonuccal, Kim Scott and Stan Grant, amongst others, speaks to this; their writing challenges the pervading ideologies of white colonial Australia and encourages readers to resist, whether in action or in thought, the Euro-Western perspective of Indigenous Australia and the impact (or assumed lack of impact) of colonial invasion.

Works by white, non-Indigenous Australian authors such as May Gibbs' *Tales of Snugglepot and Cuddlepie* (1918) and Norman Lindsay's *The Magic Pudding* (1918) are examples of how colonialism influenced early Euro-Australian literature which engages with elements of native habitats and nonhumans. These narratives prominently feature Australian nonhumans, as seen through a colonial lens. *The Magic Pudding* in particular alludes to a strongly colonialist, nationalistic message in what Christopher Kelen describes as its use of "mainly indigenous fauna … [that] behave as settlers of a new Europe, in a recently cleared landscape of smouldering stumps" (2007, 65). More recent examples that do something similar include Mem Fox's *Possum Magic* (1983) and Sheena Knowles' *Edward the Emu* (1988)—also works of children's literature which centre native Australian animals in settler-colonial stories of belonging.

In the context of earlier colonial Australian literature (writing before 1855), Elizabeth Perkins suggests that the reading of Australia's natural world from a colonial standpoint is "an assurance of future destruction", as the beholder's physical presence alone represents an invasion (1988, 139). She argues that many settlers of the time "were disturbed by an Aboriginal civilisation they could not understand" (1988, 140). Indigenous people were therefore often excluded completely from non-Indigenous stories about Australia or were portrayed problematically. Emily Potter explains that an Indigenous presence was often depicted as a haunting through "ghostly apparitions [that] refuse the forgetting of prior presence, challenging the narrative of a legitimate non-indigenous claim to place" (2019, 17). In her discussion of non-Indigenous Australian authors writing in the second half of the twentieth century, Jeanine Leane describes how these authors (such as Patrick White and Nene Gare) often marginalised Indigenous characters in their work through the use of certain motifs, such as borders and fringes which she describes as "common spaces where Aboriginality is located" (2010, 34) within these fictions. Such texts also reflect stereotypes about Indigenous family relations, employment and living situations and are, in Leane's view, typically the types of

Australian literature that students come across before reading any Indigenous Australian authors (2010, 34).

In more recent years, some modern Australian settler-colonial literature attempts to move past this detached spectatorship of the Australian natural world and the continent's Indigenous peoples, instead seeking a deeper engagement with Indigenous epistemologies and condemning the settler-colonial destruction of many native Australian habitats. Jennifer Mills' *Dyschronia* (discussed in Chap. 4) is an example of this, as is Catherine McKinnon's *Storyland* (discussed in Chap. 5), both of which condemn colonial exploitation of the natural world and explore more ecocentric ideas of kinship with the nonhuman. The increasing representation of Indigenous Australian literature within the Australian literary canon as a whole speaks to this as well, and is potentially influencing settler-colonial authors in the way that they portray non-Indigenous connections to land and the nonhuman in their work. As I will elucidate further in this and later chapters, while Indigenous understandings of the nonhuman have been largely denied in Australian settler-colonial literature, I believe that a shift is taking place that points toward an Australian literature that is more centrally engaged with Indigenous epistemologies and less concerned with settler-colonial approaches to Australian environments and the nonhuman (or, in some cases, is explicitly critical of such approaches).

Despite this shift, colonial nationalism is still influential in postcolonial Australia, which is largely dependent on the deeply ingrained colonial structures of government, economy and society. As Patrick Wolfe contends, "invasion is a structure not an event" (1999, 163); it is not something that is past, but is rather a lingering trauma that still impacts those whose lands have been invaded, ensuring that the systemic oppression of Australia's Indigenous peoples continues to this day. This reality is depicted in *The Swan Book*, with Wright demonstrating that colonisation never ceased and exists in much the same way now as it did in the early years of European invasion.[1] Deborah Bird Rose puts forward a similar perspective in the context of the Anthropocene, stating that:

... the vision of reconciliation [between Indigenous Australian peoples and non-Indigenous Australians] has been reduced to gestures that are nearly empty; many opportunities for halting the degradation of country

[1] This is demonstrated in *The Swan Book* through the continuation of Australia's Northern Territory Intervention and the subsequent containment of Indigenous people in what are effectively refugee camps.

have been thrown out, and we seem to have entered an era of maximisation of environmental pillage on the one hand, and on the other hand of increasing bureaucratic control, along with vastly expanding containment and incarceration, of Aboriginal people. (2013, 207)

As Rose argues here, the continuation of colonial attitudes towards Indigenous Australian peoples and the exploitation and degradation of stolen lands are in fact intensifying in the age of the Anthropocene. It is ill-advised to assume that "the divide between Indigenous people and settler-descended people" no longer exists—in many ways, it is now complicated by what Rose describes as the "processes that are unmaking the world that any of us ever knew" (2013, 208).

For Indigenous Australians, colonisation is an ever-present reality; Indigenous Australian author Tony Birch reflects on this in his statement that "for Indigenous people, the impact of climate change is not a future event. It has occurred in the past, and it is occurring now" (2015). If postcolonial theory is to appropriately address first-hand experiences of colonised people living within a still largely colonial nation, it is important to take into account Indigenous peoples' ongoing experiences of colonisation and the possibilities and inherent difficulties of decolonisation. Decolonising literary practices are of particular importance in the context of this chapter, especially in terms of how texts such as Wright's may be judged as magical realism (see Takolander 2016) or speculative fiction (White 2014, 145) by some critics. The novel may instead be more accurately described as an example of Indigenous realism, providing a different reality of climate change to the reader than the "stable, singular objective reality" of Western realism (Potter 2017, 245).

Evelyn Araluen expands on the usefulness of engaging with such Indigenous narratives on climate change compared to Western modes of realism that reinforce the human/nature divide by asking, "how can Western realism comprehend its own end, if it has proved incapable of recognising that which it now destroys?" (2019). The stronger emphasis on Indigenous scholarship at tertiary institutions in Australia is potentially ensuring that Australian postcolonial scholarship is becoming more responsive to these first-hand Indigenous understandings of living in a postcolonial nation. Some scholars' focus is on integrating decolonisation practices into both Indigenous studies (Nakata et al. 2012) and postcolonial studies (Swadener and Mutua 2008, 31), and such practices and understandings can also be extended to discussions of Australian literature.

Just as decolonisation seeks to challenge and ultimately eradicate Western-colonial systems of domination, ecocritical theory is often focused on how dominant Western modes of thinking—many of them implicit within the colonial project—have resulted in the wide-scale exploitation and degradation of the natural world. Travis V. Mason et al. describe ecocritical scholarship as "a conceptual model cognizant of cultural-material intersections" (2013, 3) and subsequently argue that theory stemming from ecocriticism has the ability to undermine the various dominant ideologies that impact marginalised groups (4). Such ideologies are, of course, directly relevant to the concern of postcolonial studies. Although ecocritics have attracted some critique for their emphasis on the marginalised *nonhuman* and their lack of emphasis on the marginalised *human*,[2] Mason et al. suggest that through ecocriticism's exploration of the nonhuman, the field also "reconsiders what it means to be human and the ethics that support or deny such reconsideration" (2013, 4).

In this sense, ecocritical discourse creates pathways for deeper understandings of intra-human (human–human) domination as well as interhuman (human–nonhuman) domination, whilst postcolonial theory creates opportunities for those working in ecocriticism to transform these ethical issues into activist models (Mason et al. 2013, 5). Scholars working within and close to the environmental humanities have similar opportunities to undermine such dominant, Euro-Western ideologies. Ecocritical theory, the environmental humanities and postcolonial theory are therefore critically intertwined in their emphasis on reconsidering and dismantling the problematic anthropocentrism inherent in Western perspectives of nature and the nonhuman. These various fields play a large role in this chapter's textual analysis, especially of *The Swan Book*, which addresses the relationship between the destruction of human and nonhuman habitat and Indigenous genocide.

3.2 Colonialism's Climate Crisis and the Mobile Animal in Alexis Wright's *The Swan Book*

Alexis Wright is one of Australia's most pre-eminent authors. Her novel *Carpentaria* (2006) won the prestigious Miles Franklin Award in 2007 and in 2018 Wright was awarded the Stella Prize for her non-fiction work, *Tracker* (2017), making her the only author to have ever won both prizes.

[2] For a discussion of some of these early criticisms, see Gifford (2008).

Her 2023 novel *Praiseworthy* won both the Miles Franklin and the Stella Prize in 2024. A member of the Waanyi Nation, an Indigenous Australian people based south of the Gulf of Carpentaria (Wright 2014, 342), Wright draws from Indigenous Australian concepts of Country in her writing. As described by Ben Holgate, Wright employs elements of traditional Indigenous Australian storytelling to weave a complex narrative which portrays Australia as still colonised—rather than decolonised—and its native habitats vastly altered by this colonisation (2015).

Published in 2013, *The Swan Book* is Wright's third novel and was shortlisted for the Miles Franklin Award. It imagines a future Australia rendered partly unrecognisable by climate change. European colonisation has resulted in segregated communities where Indigenous people are forced to live alongside climate refugees from other nations. As Maria Takolander describes, Wright's novel can be read as "a vision of Australia three centuries after the colonial invasion of 1788" (2016, 112). In this future Australia, the controversial 2007 Northern Territory Intervention is still in place—a policy that enabled the conservative Australian Federal Government at the time to send increased numbers of police and defence forces into remote Indigenous communities; Takolander describes this historical reference as being at the heart of Wright's novel (2016, 115). The setting of *The Swan Book* therefore represents an amplification of current social and political trends regarding the treatment of Australia's Indigenous people, whilst also offering a climate imaginary that reflects scientists' current predictions of what life will be like for humans, and nonhumans, living under the severe impacts of climate change. White describes Wright's novel as a work of "speculative fiction which use[s] the setting of a near future to show ... characters' responses to climate change" (2014, 145), but it is also a novel which depicts characters' responses to a still largely colonial Australia. It should be noted, though, that defining Wright's novel as speculative obscures the text's placement within a tradition of Indigenous realism, or "Waanyi realism", as Leane describes Wright's *Carpentaria* (2015, 161). Potter refers to *The Swan Book* as "a text by an Indigenous Australian woman that explicitly connects a climate-changed future to a colonial past within the world-altering parameters of Indigenous reality" (2017, 243). She goes on to describe how Wright's novel "writes into the future that is already our present, and as the reader quickly discerns, our past" (2017, 247).

The Swan Book describes the journey of Oblivia Ethylene, a young Indigenous Australian girl living in a refugee camp in Australia's north, surrounded by swampland and inhabited by both displaced humans and nonhumans. In the recent past, Oblivia has been gang-raped; she disappears from her community until Bella Donna of the Champions, a European climate refugee living in a dilapidated ship hull on the swamplands, rescues her. Oblivia is then sought out by Warren Finch, Australia's first Aboriginal president, who wishes to make Oblivia his wife, with or without her consent. Oblivia begins the journey south to one of Australia's cities in the custody of Warren and his three bodyguards, referred to as the "genies" (Wright 2014, 163), with the ghostly presence of Bella Donna and the Harbour Master following her on her journey (201).

Jane Gleeson-White describes *The Swan Book* as a "love story"; it is a book "about a girl's affair with the northern skies and her quest to regain sovereignty over her own brain" (2013). This quest for sovereignty is symbolic of what White describes as the stripping away of Indigenous identity and the struggle "for autonomy despite the invasion of their land and families through government control" (2014, 147). Wright describes Oblivia's "sovereignty over [her] own brain" (2014, 4) as "maintain[ing] an idea of the sovereignty of the mind, even if we haven't got sovereignty of the country or the land" (Wright and Zable 2013). Relatedly, Mead believes that Wright's novel is "a multifaceted and self-reflexive address to political and social aspects of sovereignty" (2018, 527). Themes of colonisation, dispossession and the nonhuman are therefore strongly interlinked in Wright's novel, making *The Swan Book* a meaningful example of both decolonial and environmental literature.

Throughout the narrative, Wright portrays some Indigenous Australians as working alongside the country's white colonisers; President Warren Finch is an Indigenous man, but is depicted as violent and manipulative, and someone who perpetuates the subordination of Indigenous Australians by the colonisers while presenting a pro-Indigenous façade to the public. This is evident when Warren first visits Oblivia's community and offers a small amount of government funding to improve living conditions. This is met with derisiveness by one of the community's representatives who says that "he was not interested in *loose change from Canberra*" (Wright 2014, 140; italics in original), suggesting that any past assistance from the government has resulted in little, if any, positive change. Nevertheless, Warren is not necessarily a colluder but rather represents the few life choices available to him—either accept power within the dominant systems of white colonialism or face a life in the swamplands. Wright acknowledges these complexities throughout the text, which highlights the challenging position of Indigenous politicians and advocates, such as Noel Pearson,[3] attempting to make change within a still largely colonial and often highly conservative government.

The connections Wright makes between Australia's colonisation, Indigenous dispossession and the impacts of climate change ground her novel in a realist, rather than speculative, space; what is occurring in *The Swan Book* has occurred, is currently occurring and will continue to occur into the future under the influence of colonial Australian practices and politics. Potter summarises how, far before Euro-Western philosophers began to cite the "end of nature", Indigenous ontology placed human life and culture within "wider networks of more-than-human life and power, in dynamic configurations of space and time" (2017, 246). In relation to this, Araluen comments that Western realism is limited in its ability to address ecological catastrophe because, as she states, "the end of nature, and therefore the end of externality and the limits of Western modernity, means also the end of literary forms invested in, and reifying the dominant epistemology of a dualistic relationship between nature and man" (2019). That is, realism in a Western sense privileges "an individualised, stable, singular objective reality" (Potter 2017, 245, cited in Araluen 2019), which is an unsatisfactory means of understanding climate change from outside of settler-colonial perspectives and in more ecocentric ways.

What may be viewed as a speculative exploration of the impacts of climate change and the potential for multispecies relationships between humans and nonhumans is, in fact, realist from an Indigenous Australian perspective, which in various ways challenges the Cartesian dualism inherent in Western realism, as detailed above. As a work of Indigenous realism, *The Swan Book* also puts forward the idea that for Indigenous people, environmental devastation has already taken place at the hands of the colonisers, and climate change is merely an extension of this. Kyle Powys Whyte argues this in his description of anthropogenic climate change as "an intensification of environmental change imposed on Indigenous peoples by colonialism" (2017, 153). The idea that an inextricable connection exists between humans and their surrounding environment is, as Potter claims, "hardly radical for Indigenous Australians" (2017, 246). With all of this in mind, I consider *The Swan Book* a work of Indigenous realism, rather than speculative fiction.

[3] Noel Pearson is a well-known Indigenous Australian lawyer and activist who has worked extensively in the Cape York region of Australia (Griffith Review 2021).

This near-future Australia of the novel also represents a critique of postcolonial nations in their internalisation of still-dominant colonial assumptions. Wright's clear condemnation of colonial and postcolonial systems of dominance makes clear the need for decolonisation in Australia. In this sense, a postcolonial nation such as Australia is not, and perhaps will never be, truly independent from its colonial past as long as it retains the systems of dominance—such as the subordination of Indigenous Australian people and the misuse of native habitats for settler gain—ingrained within its political and societal systems since European invasion.

It is important to note that dispossession and genocide are key themes in Wright's exploration of environmental catastrophe and its impact on the Indigenous characters of *The Swan Book*. Mudrooroo argues that Indigenous Australian literature is often directed at "the invaders of our land" and that many Indigenous writers want to "get across to as many people as possible the Indigenous predicament in Australia" (1997, 2). Wright achieves this through intense imagery of ecocatastrophe and the depiction of deep connections felt by Oblivia and other Indigenous characters with the Australian nonhuman. Wright not only utilises portrayals of the nonhuman to engage the reader with the terrifying future reality of climate change, but to describe what is really a "ghost place" (Wright 2014, 334) for Indigenous Australians, embodying the nature of colonisation and dispossession both in Wright's imagined future and now. I also argue that *The Swan Book* supports an ecocentric ethos through Oblivia's role as messenger for the nonhuman; if humans make the ethical choice to more deeply engage with the native nonhuman, then a positive shift towards a less mechanistic perspective of the more-than-human other may be possible.

Multispecies theory is a useful lens through which to analyse the human–nonhuman relations depicted in Wright's work, and the implications of such relations in our understanding of the modern environmental crisis from a postcolonial perspective. Many of Australia's species are part of what Isabelle Stengers terms "the political fray" (2010, 20, cited in Ogden et al. 2013, 15)—they have become entangled in the politics of conservation (Ogden et al. 2013, 15), as well as the fight to retain settler-colonial identities tied to certain animals (such as the protection extended to non-native, wild horses, or Brumbies, and farmers' desire to eliminate the native Dingo). The fate of both native and non-native nonhumans often lies in the hands of politicians deciding on environmental policy, and the agential rights of these beings can be brought into question or ignored through such policy. It is difficult, if not impossible, to preference one nonhuman's agency over another. This is particularly relevant in the case of Australia's wild Brumby populations, who are simultaneously viewed as an invasive species and a national icon (Nimmo and Miller 2007) and whose agency within their Australian habitat directly conflicts with that of the native species also inhabiting this space.

These multispecies considerations within Australia's ecological and political reality are reflected in Wright's novel. The reader is introduced early to the omnipresence of extinction in this near-future world; the first chapter following the prelude describes how "in every neck of the woods people … talked the language of extinction" (Wright 2014, 6). The constant movement of the black swans throughout the narrative hints at the very real issue of native species translocation and population decline due to anthropogenic climate change and habitat destruction at the hands of settler-colonisers. For many of Australia's native species, this spells extinction or, at a minimum, endangerment.

3.3 Breaching the Borders Between Bird and Girl

Wright's portrayal of the intermeshing stories told by European climate refugee Bella Donna and experienced by Oblivia highlights the importance of overcoming the borders between humans and nonhumans, and the nature/culture dualisms inherent in settler-colonial understandings of the nonhuman. Bella Donna's first encounter with Australia includes the smell of bushfires, the sight of "*gidgee kadawala* woodlands and bloodwood *corymbia capricornia*" (Wright 2014, 32; italics in original); the use of both the Indigenous Waanyi language and Western taxonomy within the one sentence reveals a layered perspective of native Australian habitats. Similarly, Oblivia hears Bella Donna's stories of white swans but she also observes how Bella Donna draws the black swans to the swamplands: "the old black swans had heard her voice running along streams of dust floating in the breezes" (Wright 2014, 65). This suggests that the way non-Indigenous Australians consider the native nonhuman often stems from their perceptions and relationships with the non-native animals and plants that are more familiar to them. Directing people to connect with native nonhumans through a decolonised lens is therefore important if Australia's native species and habitats are to be better understood.

It is subsequently significant that Wright's novel opens with a discussion of belonging. The unknown narrator (who the reader later learns to be Oblivia) describes their inner battle with a virus: a dichotomy between the body and an invader which raises questions regarding colonisation and the nonhuman. The narrator tells the virus that they

...felt more at home with the cool air flowing on my face from a wild Whistling Swan's easy wings sweeping over snow-capped mountains in its grand migration across continents, than in those vast ghostly terrains of indescribable beauty that have given me no joy. (Wright 2014, 5)

This is the first indication that this novel explores the binary of the native versus the non-native; the "snow-capped mountains" and "a wild Whistling Swan" symbolising the colonising Europeans and the "vast ghostly terrains", the quintessentially native Australian habitats of desert and shrub. Described as "a tinder-dry nimbus" (Wright 2014, 5), it is this latter habitat which the narrator wishes to connect with; a place where they "once felt a sense of belonging" (5). This opening chapter sets the scene for a narrative that seeks to shift the reader's understanding of an anthropocentric Euro-Western nation to an ecocentric Indigenous Australia.

Indigenous storytelling and storytelling in general are central to Wright's work. Arnaud Barras suggests that

Wright's novel makes clear that the field of Australian literature cannot be located precisely within the borders of the continent, but has to be understood transculturally and intertextually *in relation to the world*. (2015, 1; italics in original)

This, too, can be extended to not just the field of Australian literature, but Australian literary ecofiction. In Wright's novel, the connection felt between Oblivia Ethylene and the swan stories, both new and old, demonstrate how relationality affects individual perception; it is not just this relation to the world, as Barras describes, but also a relation to other species that expands the borders of Australian ecofiction to encompass both human and more-than-human experiences. Barras explains how *The Swan Book* "...invites the reader to reflect on the production and reception of stories" (2015, 3). Whilst the stories told to Oblivia by Aunty Bella Donna and the stories Oblivia later tells herself include largely anthropocentric perceptions of the animal, Wright's work resonates with Maurice Merleau-Ponty's notion of "the deep world of untamed perception" that exists between the human and animal (1973, 118), encouraging the reader to consider other, nonhuman stories instead of solely the human. Returning to Haraway's discussion of significant otherness and the nature/culture divide, it is important to note that in the context of the Indigenous perspectives present in Wright's novel, there is an understanding by Oblivia that nature and culture are irrevocably intermeshed. Neel Ahuja argues that these perspectives are becoming more prominent within species studies transnationally, stating that "recent scholarship at the intersection of postcolonial studies, ethnic studies, and species studies acknowledges links between species, race, and transnational power structures" (2014, 227). The multispecies relationships presented in *The Swan Book* through Oblivia, Bella Donna, Warren Finch and others strongly depict the complexities of these intersections, highlighting both the specifically Australian *and* transnational concerns of acknowledging Indigenous epistemologies in a space still dominated by colonial assumptions.

Significantly, *The Swan Book* engages with stories from all around the world—not just those of Indigenous Australia. This is central to Wright's narrative as it instigates a dialogue between Euro-Western and Indigenous Australian perspectives of nature whilst simultaneously creating a sense of universality in relation to human stories of nature. Gleeson-White argues that this is part of the novel's mission to create an "architecture of the mind", one that "teems with songs, stories, images and fragments from cultures across the planet" (2013). This intermeshing of stories invites the reader into the chaos of Oblivia's psychology, emphasising not just her own experience as a dispossessed Indigenous Australian person, but the experiences of so many Indigenous peoples living in postcolonial nations worldwide in a time of immense environmental change.

Oblivia's specific connection to the black swan, and the stories she tells and is told about them, contrasts the nonhuman which is native with the nonhuman which is non-native or invasive, highlighting the similarity between the invading colonisers and invasive species. The term "black swan" has come to be known as a way to describe an extremely unlikely event, based on European encounters with the black swan of Australia despite the previous assertion that only white swans existed. Known officially as "The Black Swan", or Black Swan theory, and developed by Nassim Nicholas Taleb, this theory "argues that too often we ignore the possibility that the world around us might change dramatically in the blink of an eye" (Lybeck 2017, 9). From an Indigenous Australian perspective, though, the black swan represents the norm—a species that has existed amongst Indigenous Australians for thousands of years—whilst white swans symbolise the catastrophic event of Australia's colonisation and subsequent environmental devastation. Although a theory often applied to economics and finance, it can be argued that "The Black Swan" and its proponents uphold this Western colonial idea that the discovery of something *foreign* (that is so different from the *common*) is revolutionary for the discoverers, ignoring the impact of the said discovery on those who have lived alongside this being all along.[4]

Oblivia's relationship with the black swans suggests that she is challenging the dominant influence of colonisation within a society that still largely accepts white swans over black. Bella Donna's stories are a resounding reminder of this, as she describes: "Our swan's feathers shining like gold … and the pearl that hung from its neck, the size of a huge ball, shining in the moonlight" (Wright 2014, 31). This and other stories told by Bella Donna to Oblivia reveal Bella Donna's European heritage, the huge pearl around the white swan's neck and the swan's bright feathers representing a blinding whiteness—one that obscures a deeper, more intimate connection between Indigenous peoples and their Country. It is depicted as unusual when Bella Donna finally deviates from her stories about white swans to tell Oblivia, "A black swan flies slowly across the country, holding a sliver of bone in its beak" (Wright 2014, 44). The white bone represents the new dominance of the black swan over white colonialism, signalling what Barras describes as "the Aboriginal people and their struggle for freedom and emancipation in the aftermaths of colonisation" (2015, 6).

Whilst most of Wright's novel focalises Oblivia's relationship with the black swan, Warren Finch—true to his surname—also maintains a deep connection to native bird species. The brolga, an iconic Australian animal, is seen by the younger Warren undertaking its "supreme ceremony of Country" (Wright 2014, 111), representing a cultural parallel to Indigenous Australian traditions. Wright portrays these birds as capable of enacting cultural behaviours, "the teeming brolgas searching across the landscape for the distant music of the night-time minstrels" (2014, 111). Warren is depicted as existing amongst this flock of birds, inhabiting their space rather than them inhabiting his. This deep relationship with the nonhuman which stems from Warren's childhood symbolises what Warren often denies throughout his presidency—a profound and interminable connection to Country and his Indigenous past, despite his colonial present.

[4]This relates to Ross Gibson's discussion of Australia as an inverse place that is "oddly doubled" from the point of view of Westerners, who "recognize themselves there at the same time as they encounter an alluringly exotic and perverse entity" (1992, x).

The black swan, too, plays a role in Warren's childhood, when he comes across an injured bird as a boy. Warren notices the swan's struggle, realising that they are "tangled in a bundle of fishing line knotted in the roots and tree branches" (Wright 2014, 114). His choice not to help them, to watch the bird "pulled helplessly through the water [as] a wall of floodwater exploded through the piles of sticks…" (Wright 2014, 114), anticipates Warren's future powerlessness as president to make real change within a white colonial government—change for both Indigenous Australians and for the native nonhumans impacted by anthropogenic climate change. Like the sovereignty of Australia's Indigenous people pulled beneath the overpowering waters of colonialism, the swan is drowning: "…drowning for Warren Finch, and all the boy saw were pictures of Aboriginal spirits with halos of light, just like Van Gogh had painted" (Wright 2014, 114). Such imagery is indicative of the Indigenous struggle against an ever-present colonialism. The reference to the European Vincent van Gogh represents how Indigenous activism and the environmental crisis are too often viewed through a Euro-Western lens. However, the symbology of Warren's interaction with the swan also represents his sadistic treatment of Oblivia (his abduction of her and her imprisonment within Warren's urban "hell" (Wright 2014, 215)) and his alignment with Western instrumentalism, as he ponders whether the seemingly incompetent "Aboriginal Governmental Nation" of the swamp was "even worth saving at all" (Wright 2014, 148).

Before the swan drowns, Wright describes the bird as "content to stay where it was…" (2014, 113)—a forewarning that Warren's leadership represents an acceptance of colonialist power structures as the waters pull him and other Indigenous people further under. This posits that decolonising perspectives represent a more constructive critique of the still largely colonial power structures that exist in Australia. Warren's relationship with the nonhuman also reveals him to be a more complex character than the reader is first introduced to. Through his villainous actions to abduct and forcibly marry Oblivia, he is in many ways acting for the colonial state, but he is also shown to be grappling with his own understanding of and connection to Country within the confines of the colonial systems of dominance that made him president. The swan and, by extension, Warren should *not* be content to stay where they are, within this colonial framework that continues to pursue and uphold Indigenous dispossession, but should rather embrace decolonising practices in an effort to oppose the systemic oppression of Indigenous Australian people.

It is through Oblivia, whom I define as a messenger, that Wright reveals the true nature of the human–nonhuman relationship. Oblivia represents the possible when it comes to how humankind, more specifically Australians, can learn to understand and connect with native nonhumans on a deeply personal level. This relationship with the native nonhuman extends beyond the simple act of viewing nature; it is instead a meaningful connection between kin in a future society where both the human and nonhuman are threatened with extinction. The idea of the messenger suggests that humankind requires a special connection made through individuals in order to access the world of the nonhuman.

In the case of *The Swan Book*, Oblivia's Indigeneity emphasises both the unwillingness and the inability of non-Indigenous Australians to dismantle Australia's colonial systems of domination and connect with the land and the nonhuman inhabitants that they have taken possession of as colonisers. Oblivia's name is testament to this and is described aptly by Takolander:

> While Oblivia's chemical surname highlights the parallel between the destruction of a people and the destruction of their land, her first name also draws attention to the temptation of historical forgetting. (2016, 114)

Through this, Wright focuses the reader's attention on the obvious and yet largely accepted actions of various Australian governments to not only continue the subordination of Indigenous Australians through extreme legislation and environmental destruction but to forget the historical precedents which have paved the way for current practices.

Notably, Oblivia is treated as subhuman by other Indigenous Australians, such as Warren Finch, and the white colonial system in which she is trapped. Before Warren meets Oblivia, he already refers to her possessively as his "wife" (Wright 2014, 139) and Oblivia herself does not speak, making her unable to respond to Warren's claim. This inability to speak in some ways aligns her with the nonhuman and she is ostracised and punished for this connection, suggesting that an ecocentric relationship with the natural world is unsustainable and out of place in a colonial nation. Her role as messenger also represents a deep Indigenous connection to Country—something else that she is potentially being punished for within a society seeking to eradicate its Indigenous history. Oblivia's symbiotic relationship with the "giant eucalyptus tree" (Wright 2014, 7) in which she sleeps is also emblematic of this connection to Country, highlighting her ability to communicate with the nonhuman:

> Locked in the world of sleep, only the little girl's fingers were constantly moving…. She was writing stanzas in ancient symbols wherever she

could touch … it was either the older language coming to birth again instinctively, or through some strange coincidence, the fingers of the unconscious child forming words that resembled the twittering of bird song speaking about the daylight: but the little girl could not understand the old ghost language of warbling and chortling remembered by the ancient river gum. (Wright 2014, 7–8)

This passage also emphasises the degradation of this relationship with Country due to colonisation. As the final sentence implies, Oblivia is limited in her ability to understand this particular nonhuman; its language is a "ghost language" (Wright 2014, 8), something that she cannot fully understand or that has been lost.

Like the character of Amedea in Bobis' *Locust Girl*, Oblivia does not follow a straight path. Whilst Warren is a firm believer in destiny, having been told that Oblivia is "[h]is promised one" (Wright 2014, 93), Oblivia is meanwhile depicted as someone who is pushed and pulled by others into pursuing a certain future. In addition to being kidnapped by Warren and the genies, she is emotionally torn between the opinions of the ghosts of Bella Donna and the Harbour Master. The Harbour Master, commenting on Oblivia's relationship with Warren Finch, tells her "*You better lay low … if you are going to keep hanging around with that idiot*" (Wright 2014, 203; italics in original), whilst Bella Donna warns Oblivia that Warren "would end up killing [her]" (Wright 2014, 201). Oblivia's fraught relationship with the ghosts symbolises what van Dooren and Rose describe as an undoing of "any sense of a firm border between the living and the nonliving" (2016, 87), reinforcing the idea that Oblivia plays the role of messenger between the reader and nonhuman, although this comes at the cost of having to engage with multiple perspectives rather than hers alone. Her deep connection to the native black swan is what eventually leads Oblivia back to the swamplands, along with the swans themselves. While sheltering a cygnet in her clothes, she travels unnoticed amongst a crowd of refugees on her way back to the swamp and the "loud beating wings" of "the cloud of swans" (Wright 2014, 310) keep her hidden on her journey, suggesting that her unsettling of the human–nonhuman border guides her away from the settler-colonial authorities seeking to control her.

However, it should be noted that the above reading is informed by the multispecies writings of two non-Indigenous scholars, van Dooren and Rose, and that there is more complexity to be understood here that extends beyond multispecies studies. For example, Oblivia's discourse with the ghosts also reflects Indigenous ontology regarding externality, in that human culture does not exist externally to nature and the nonhuman. Potter summarises this as follows:

Being always in relationship, always enrolled, is the pre-condition of Indigenous ontology, in which human life and culture [are] inextricable from wider networks of more-than-human life and power, in dynamic configurations of space and time. (2017, 246)

This can be linked to the Indigenous concept of Country, as described by Aunty Joan Tranter and the Gay'wu Group of Women in the previous chapter: the boundaries between humans, nonhuman forces and other living beings understood by settler-colonisers are absent from these Indigenous understandings.

It is significant that Oblivia is an Indigenous child who is portrayed as being rescued by the white Bella Donna. Gang-raped by a group of Indigenous boys "who thought they were men" (Wright 2014, 83), Oblivia's vulnerability and child-like nature symbolise the figure of what Honni van Rijswijk, amongst others, describes as the "abused Aboriginal child": a significant myth-making figure that has a played a major role in white Australia's treatment of Indigenous people (2014, 118). It is this figure which was primarily at the centre of the Australian Federal Government's "Northern Territory National Emergency Response Act 2007", also known as the Northern Territory Intervention (hereafter referred to as "the Intervention"), as well as the enduring practices of Indigenous child removal by the state. Van Rijswijk argues that *The Swan Book* challenges this image of the "abused Aboriginal child" and is, in effect, a "counter-archival practice that … might challenge law's archive by making visible its silences and omissions" (2014, 120). In this context, "law's archive" is what the Intervention, and the following "Stronger Futures in the Northern Territory Act 2012", instigated in the Northern Territory; the government's seizure of community land leases, the deployment of the army into Indigenous communities, the restriction of welfare benefits to people in these communities, and an added police presence are just some of the extremist actions taken by the government to combat issues of child abuse in these places (2014, 118).

As exemplified by Wright's work, Indigenous Australian literary practice plays a significant role in decrying these policies and their consequences for Indigenous people. Bruce McGuinness and Dennis Walker contend that

Aboriginal writers have a responsibility here … to take that message not only to white people but to Aboriginal people as well, so that we can foster in our own communities a very important concept. That concept is that we are going to survive, we are going to have to do it as a community, we are going to have to do it as a nation and not as individuals. (1985, 49–50)

Wright puts forward this view in her work, arguing throughout the narrative, but most importantly in its conclusion, that Australia's Indigenous peoples must take the natural environment back into their care not only for the sake of the nonhumans inhabiting it but for the sake of Indigenous survival in a still largely colonial nation. This message is intimately entwined with descriptions of the nonhuman, with Wright likening of Oblivia's mourning for Country in the novel's epilogue to "a sigh of a moth extending out over the landscape, or a whisper from the scrub ancestor catching a little stick falling from a dead tree…" (Wright 2014, 334). As this implies, colonisation and ecocatastrophe are linked. Ogden et al. describe how ecology is inherently politicised, stating that

[r]ain, cloud, and temperate forests, savannas, oceans, and polar expanses—all are sites of contest, development schemes, colonial and postcolonial displacements, resource wars, and capitalist expansion. (2013, 15)

The dispossession enacted against Indigenous Australian peoples is therefore not solely a human issue, but one which has resulted in, and is still resulting in, the destruction of Australia's nonhuman species and habitats. The decolonisation that Wright argues for in her novel, depicted through Oblivia's embrace of the black swan and the hopeful return of "*Bujimala*, the Rainbow Serpent" (Wright 2014, 334; italics in original) at the novel's conclusion, suggests that a rejection of colonial control may lead to Australian nonhumans being given the opportunity to, as Erika Cudworth describes, "flourish" (2005, 7).

Wright's narrative portrays Australia as in dire need of decolonisation. In the case of the black swan, Oblivia's connection to the species represents not only Indigenous ties to the natural world but also a hopeful message of Indigenous sovereignty, centralising Indigenous perspectives on the future of Australia as a decolonised nation in a novel about climate change. In the penultimate chapter of Wright's novel, Oblivia perceives that "[s]omewhere in this landscape, swans were stirring" (Wright 2014, 303). The sky above her becomes "blanketed by black swans in the cold night" (Wright 2014, 303), symbolising an Indigenous uprising against the colonial powers responsible for the displacement of both the native nonhuman and the Indigenous peoples across the country. The swans are returning home to the swamplands, and Oblivia realises that "she had found her swans" (Wright 2014, 303). The entirety of this passage in the novel is significant, strongly suggesting a deep link between connection to the nonhuman and connection to Country, but also between ecological empowerment and the fight for decolonisation:

They had found each other's heartbeat, the pulse humming through the land from one to the other; like the sound of distant clap sticks beating through ceremony, connecting together the spirits, people and place of all times into one. These were her swans from the swamp. There was no going back. She would follow them. They were heading north, on the way home. (Wright 2014, 303)

This hopefulness is partly challenged when the epilogue of the novel describes Oblivia wandering the swan-less swamplands, now "too hot and dry" for the species to live here (Wright 2014, 334). It is a "spirit-broken place … where the Army owned everything" and where Oblivia is heard crying "*kayi, kayi kala-wurru nganyi, your country is calling out for you*" (Wright 2014, 334; italics in original): a lament for her people to return home, and a call to action for decolonisation that could be perceived as retaining some of the hopefulness represented through Oblivia's relationship with the swans (the finding of "each other's heartbeat" (Wright 2014, 303)). These final passages perhaps do therefore allude to a hope that decolonisation is achievable for Indigenous Australian peoples, and Australia more broadly; after all, it is stated that the "swans might come back" (Wright 2014, 334) and reclaim the swamp from which climate change and, by extension, colonisation had banished them.

3.4 Multispecies Migration in Merlinda Bobis' *Locust Girl*

Bobis' *Locust Girl* tells the dystopian story of Amedea, also known as Beena, a young woman attempting to understand the events of her past, when her family and village were annihilated by the Five Kingdoms—a largely faceless authority who exist behind a giant border, created by them to restrict Amedea's people—the "strays"—to the arid wastelands, while the privileged live in the lush, green lands of the Kingdoms. At the beginning of the novel, Amedea survives the bombing of her village but is buried in the sands below it. She emerges ten years later with a locust embedded in her forehead, pulled from the sands by Beenabe—another young woman searching for her family and village. Throughout the narrative, Amedea's locust sings mysterious verses, often landing her in trouble with fellow strays and the authorities because singing is outlawed. Amedea travels with Beenabe, meeting other strays who teach her about the all-encompassing power of the Five Kingdoms, who at their best force the strays to line up for limited food and water rations and at their worst harvest the strays' organs and body parts and bomb whole villages at any sign of disobedience. Amedea is eventually separated from Beenabe, but meets her again in the latter chapters when Amedea herself makes it behind the border, into the land of the Five Kingdoms, who she learns have been punishing the strays as descendants of previous nations who supposedly destroyed the environment through their own exploitative practices and bombing campaigns. After failing to escape the attention of the authorities behind the border, Beenabe is murdered and Amedea is forced to return to the devastated lands outside of the Kingdoms, where she is reunited with the other strays.

The subtitle of the novel ("a lovesong") reflects the emphasis on song and music throughout the novel, but begs the question of who the lovesong is sung by and who it is for. This is a consistent tension throughout the narrative, as it is often unclear whether it is Amedea, the locust or both of them together that sing the haunting verses in response to different events. In Bobis' novel, the connection between human and animal is presented as an intricate dance—one that reflects the constantly shifting forms of responsiveness, reciprocity and accountability that challenge the dominant "only-human stories" (van Dooren et al. 2016, 2–3) that have shaped the anthropocentric attitudes of Western modernity. This is evidenced in the following passage describing the locust and Amedea's first meeting under the bombed remains of Amedea's village:

The nibbling thing was the last of the grey underground creatures with once blue eyes and whiskers…. It was dislodged from its nest by the explosion and had fallen into a deeper hole with me…. It nibbled at me, thinking I was a stone blocking its way…. Then it grew tired. It nibbled its way under my forehead and there slept my ten-year sleep. (Bobis 2015, 9)

As in *The Swan Book*, sleep and dreams are a recurrent theme in *Locust Girl*; it is often in these in-between places of consciousness that the protagonist comes to understand the animal better. While Amedea's and the locust's shared experiences in wakefulness are often convoluted and unintelligible to Amedea ("I couldn't make sense of the song that I heard in my head" (Bobis 2015, 24)), their shared sleep and dreams articulate a singularity of mind that surpasses the confusion of their waking world:

We listen to the other's dreams
In the other's skin—once a locust
And a girl, then a locust girl
Dreaming a single dream (Bobis 2015, 9)

It is significant that the reader is first introduced to the agency of the locust through dreams, rather than through their actions in the waking world. The locust dreams of Amedea just as Amedea dreams of the locust; each listens to the other.

While it is perhaps easier to read the word "locust" in "locust girl" as an adjective rather than a noun, the setting up of the title in the above passage suggests that both are in fact nouns—the locust becomes part of the girl in the same way that the girl becomes part of the locust. "Girl locust" would perhaps be just as fitting here, then. The locust communicates on behalf of Amedea and, by the end of the novel, Amedea's mode of communication is both her own and one that is merged with the locust: "…it made me sing a new song. I sing it to you now" (Bobis 2015, 179). It is perhaps an over-simplification, then, to say that Amedea is a messenger for the nonhuman in the same way that Oblivia is, because the locust is often who speaks for Amedea. This is a two-way mode of agency and communication, the locust not simply a symbolic extension of Amedea's human mode of expression.

It would be easy to take the novel's politics as a purely human politics, with the animal and the nonhuman functioning as metaphors of intra-human oppression and human displacement under colonialism, authoritarianism and associated ecological collapse. After all, the novel can be read as a political allegory about border policing and the oppression and exploitation of migrants and asylum seekers in the twenty-first century. However, I believe that the politics of this text extend beyond the human and are consequently all the more potent. As I will go on to describe, Bobis' consideration of the idea of reciprocity—and the colonial-capitalist misinterpretation of this concept—is particularly relevant in this regard.

Bobis draws attention to the linguistic constraints of the Western novel form, particularly how the animal is often used as a metaphor for solely human struggles. The locust in *Locust Girl* is more than a metaphor for action against intra-human oppression, seeming to play a role that goes beyond the purely metaphoric. The fact that singing is outlawed implies that there is no room in this colonial system of discrete nations and borders for any kind of messy, entangled expression through singing and music or the overlapping relationships between the human and animal. As

I will explore further below, the laws of the Five Kingdoms maintain a separation between human and animal in an attempt to purify nature as a protected, inaccessible resource. This appears possible within the borders, but is less possible outside, in the wastelands where Amedea comes from. Bobis approaches a decolonial ethic in her narrative by offering the idea that what is possible outside of the borders is resistance via human–animal connection and communication—after all, “it’s not us who are singing, so we’re not breaking the law” (Bobis 2015, 92), says Rirean, a waster who joins others in listening to the locust’s songs as if they were prophecies. This co-singing could be seen as a “passionate immersion” (van Dooren et al. 2016, 6) in diverse lifeways beyond the human that afford Amedea an understanding of the impacts of the Five Kingdoms’ scorched-earth policy on both the human and nonhuman.

Despite the emphasis on the animal here, we must still consider the human politics of this text—borders, refugees, authoritarianism—in order to understand how they link with more-than-human concerns. Returning to Belcourt’s work, he argues for a stronger focus on spatiality “to get at the ways in which animals are emplaced in a constrained set of zones of living and dying” (2020, 21). Addressing spatial issues in this context means understanding and challenging the settler imaginary’s assertions that “animals exist … as geographic subjects, enervated and trapped” (2020, 21). Literary representations of both borders and the animal together are therefore a good place to start here.

Before the reader even begins the story of Amedea, they are introduced to the novel’s emphasis on borders in Bobis’ dedication:

For those walking to the border for dear life
And those guarding the border for dear life (2015).

This not only sets up the focus on the affective quality of borders on the lives and livelihoods of those engaged with them; it also hints at Bobis’ critical approach to dualisms in the narrative. Borders facilitate such separation—Bobis’ work does not deny their existence nor their affect, but instead seeks to understand what living around and, later, dismantling such borders may or may not look like to different groups.

Bobis’ fiction is especially concerned with the political implications of different approaches to spatiality. In his analysis of Bobis’ short story collection *White Turtle* (1999), Patrick West describes a form of spatiality reflected in Bobis’ storytelling that exists “between the divisive poles of (literary) metaphoricity and so-called real-world space” (2024, xv–xvi). The eponymous story of the collection, “White Turtle”, is particularly attuned to this, West stating that "it instigates an intensively spatial … investigation into the relations and overlappings of space with linguistic and non-linguistic materialities" (2024, 166). *Locust Girl* arguably achieves something similar with its interplay between the language of the locust and the language of Amedea, who, like Oblivia, is often silent throughout the novel. After emerging from her burial beneath the desert sands, Amedea struggles to understand her surroundings and articulate this struggle to those around her. The locust seems to take on this job for her, communicating to others through verse but also with Amedea herself, assisting her in her journey across the arid desert space and then through the lush green landscapes of the Kingdoms. This interplay is partly a non-linguistic materiality—the locust is a real animal, embedded in Amedea's head, that has a presence beyond its singing ability. There is an allusion to the animal controlling or influencing her physical movements as well, Amedea often finding herself in a new place with little to no knowledge of how she arrived there: "I found myself looking at half a bowl" (Bobis 2015, 30), Amedea confusedly reflects during her journey through the wastes with Beenabe, not understanding how or why she came to be observing this object. What might be termed a *linguistic* materiality, though, is the way in which the locust's singing, and not just its presence, assists Amedea in articulating and understanding her own relationship to spatiality through movement: "It began as simultaneous notes, like a burst of harmonies that slowly tapered into three notes ultimately parting ways but all pushing our feet towards the rock. Soon the three notes became two, echoing the weary rise and fall of our soles" (Bobis 2015, 37).

Amedea's perception of "the border" is also altered by the locust, as she is encouraged to move through different landscapes in companionship with the animal. On the "outside" of the border—in the wastelands, where the strays reside—the border is usually seen as the horizon, a blurry, indistinct line that seems constantly out of reach: "that edge, that long line, which she wanted to cross, was always beyond home, wasn't it? The sun always sank into that line, but now it seemed to be sinking into where she came from" (Bobis 2015, 15). Comparatively, when Amedea moves into the land of the Five Kingdoms, the border is barely spoken of and it is unclear if it is even visible to those privileged few that live within it. This is despite it being mentioned consistently in the Honourable Head's (the leader of the Five Kingdoms) political rhetoric, in his urgings that "there is peace if the border is protected. There is purity if all keep to their own place under their own sky" (Bobis 2015, 121).

Bobis deconstructs the human/animal binary by attending to the border between them. This border is more than a dividing line, though—like the locust, it has an agency unto itself, an edge or a verge that contains multitudes within it, and is itself not wholly good or wholly bad. One of the songs sung by the locust, and repeated throughout the narrative, is the following:

The edge is a line, oh how lovely
It will stretch your eye
The edge is a line, oh how sharp
It will cut your feet (Bobis 2015, 22)

Here, the edge—which one might interpret as the border—is both enticing and dangerous. It is a site of coming together, of "new possible paths and interactions" (Crosby et al. 2014, np), but is it also a site of divergent conflict—having to walk away from the edge, of being separated by a border that designates one side as good and the other as bad. For much of the novel, crossing the edge or border feels wholly out of reach for Amedea: "that edge, that long line … was always beyond home" (Bobis 2015, 15).

This separation is represented by the categories of "wasters"—those who exist in the wastes, outside of the border, also known as the strays—and "carers"—the privileged group that resides within the borders of the Five Kingdoms. Amedea is a "waster", not a "carer", but she has a closer relationship with the animal, one that seeps into Amedea's personhood just as her personhood seeps into the animal. This is exemplified by the locust's whirring form of communication and its transformation through a kind of singing:

The whirring grew louder, boring into her ear, urging her to her feet…. The whirring was like a singing now. It had highs and lows, and meandered through a scale. After a while it grew melodic, small notes in a wordless ditty. (Bobis 2015, 16–17)

This is a vastly different relationship to the false, idealistic "purity" imposed onto the animal by the Five Kingdoms (Bobis 2015, 149). In reality, the animal seeps into everything, though, and is largely immune to the stringent categories and borders enforced by the Five Kingdoms. Such seeping may seem like a miniscule effect in the broader context of bombs, genocide and mass displacement, but it is immensely powerful. As Steve Mentz asserts, "seeping names an ecological truth that all borders must be crossed and all boundaries spanned" (2017, 282).

To care is to engage in reciprocity—an idea which is foreign to those who ironically describe themselves as the "carers". The Honourable Head of the Five Kingdoms prioritises "purity" in the earth. Those who live within the border are voyeurs more than they are carers: observers of a "pure", manicured nature which they claim to have cultivated from the ruins of the fires supposedly caused by the "wasters"—Amedea's ancestors, those who now live outside of the border. As the Honourable Head explains to Amedea, "I purified the earth and our need for it. I controlled the seasons. Now there is a season for feasting, then for resting, and a season for just looking" (Bobis 2015, 150). He mistakes this for reciprocity, telling Amedea that "... equality is deserved only by those who are willing to care for these gifts. Gift giving is reciprocal" (Bobis 2015, 150). Reciprocity is more than just an exchange, though—it is not capitalism. True reciprocity cannot exist without a sense of kinship; without messy, imperfect human–nonhuman relations; without a gift being offered by the nonhuman themself rather than the human assuming that a resource is ripe for the plucking.

In *Braiding Sweetgrass*, Robin Wall Kimmerer recognises this key difference between transactionality and an ongoing relationship of reciprocity in her own relationship with the wild strawberries picked during her childhood, compared to those that are store-bought. Kimmerer elucidates this using the example of a gift of socks versus socks purchased at a store:

> ... I have no *inherent* obligation to those [store-bought] socks as a commodity, as private property. There is no bond beyond the politely exchanged "thank yous" with the clerk. I have paid for them and our reciprocity ended the minute I handed her the money.... But what if those very same socks ... were knitted by my grandmother and given to me as gift? That changes everything. A gift creates an ongoing relationship. I will write a thank-you note. I will take good care of them and if I am a very gracious grandchild I'll wear them when she visits even if I don't like them.... Wild strawberries fit the definition of gift, but grocery store berries do not. It's the relationship between producer and consumer that changes everything. (2020, 26)

For Kimmerer, "gifts from the earth or from each other establish a particular relationship, an obligation of sorts to give, to receive, and to reciprocate" (2020, 25). Drawing on Kimmerer's thoughts on practicing reciprocity through restoring sweetgrass, Catriona Sandilands describes this kind of "restorative ecological relationship" as "a process of attentive intertwining ... that ... defies the compulsion of capitalist accumulation" (2017, 27).

Amedea's own musings on reciprocity evoke a similar sentiment, revealing the problems with framing reciprocity as a trade, rather than an act of gifting:

So is this what love means to you, Beenabe? The love taken is equal to the love made? But must love be reciprocal, for it to be love? You can't take it if you can't make it? Must love follow the law of the market? Like in the ration lines? Your last crockery for a fistful of seeds? Your heart for a jug of water? (Bobis 2015, 159)

Because of this maintained separation between the human and nonhuman within the borders of the Five Kingdoms, there is a notable lack of deep engagement or kinship with the more-than-human that goes beyond an observer/observed, user/product binary. Following the Honourable Head's explanation to Amedea of how he formed this new "pure" earth through separation, Amedea notices as he tries to reach out to a nearby animal that the animal moves to hide behind some grass. "I only saw a flash of black and brown" (Bobis 2015, 150), Amedea notes, and the Honourable Head later explains that "the animals are not shy but mortally afraid of our want" (Bobis 2015, 151).

This is a place, then, where human–animal interactions appear almost non-existent, apart from the animals' use as food and their terrified avoidance of any human—including the "carers". Of course, the reader is expected to recognise the facetiousness of this label based on what they know about the Five Kingdoms and their treatment of those who live in the wastelands beyond the borders. Through this passage, Bobis challenges colonial wilderness theory which has informed many white environmentalist ideologies that emerged in the latter half of the twentieth century and still hold great power today—that is, in order to protect or "care for" nature, we must lock it up, maintaining the separation between nature/culture, human/nonhuman, that reinscribes colonial-capitalist ideologies of "nature" as a resource that can only be "healthy" in the absence of humans. As the Honourable Head tells Amedea, "We can't worry nature with all of our to-ing and fro-ing. This is how to preserve it. We can't soil it every day with our want. We must return it to its purity, so we too are returned to ourselves" (Bobis 2015, 149).

The Five Kingdoms' ideology of supposed "reciprocity" also informs their treatment of the wasters as a literal source for meat—most of the animals within the Five Kingdoms are protected from such a fate, but there are allusions to a black market for human body parts in which the healthier wasters are picked out of the ration lines and their limbs and organs harvested: "eyes here, last good leg there, maybe that hand with the ring" (Bobis 2015, 80). This is framed as a trade between the harvester and harvestee, as a way for the harvestee to repay the authorities for their good health, no matter how involuntary the payment might be: "rewards were reciprocal" and the more valuable body parts are "traded across the border" (Bobis 2015, 80). Here, the Five Kingdoms use humans' material animality against the wasters, demonstrated by this trade of material body parts but also by the Honourable Head's enforced monochrome palate in the wasters' environment: "all must blend with their habitat as animals would", he believes (Bobis 2015, 11).

Such rhetoric reveals the difficulty in approaching the human from an ecological and material standpoint. On the one hand, Amedea and her people are empowered by her relationship with the locust. In turn, no doubt, the locust is similarly empowered, just as fed up with life in the wastes as humans seem to be. On the other hand, though, the authority of the Five Kingdoms exploits the wasters' material forms, harvesting body parts, maintaining control of them through limited sustenance and water that would otherwise make them healthier and stronger, and, of course, punishing Amedea for her relational, material embodiment with a nonhuman other, the locust.

3.5 Conclusion

The relationships between human and nonhuman depicted in *The Swan Book* and *Locust Girl* signify an attempt to reconcile not just with the nonhuman, but with the ineptitude of many in the West to confront the environmental damage done at the hands of humans and, in many cases, colonisers. Both Oblivia and Amedea are punished for their special connection to the nonhuman animal, representing the dominance of the prevailing settler-colonial, anthropocentric attitude towards more-than-human life-worlds, even in times of eco-crisis. The protagonists' attempts to overcome such punishments symbolise a potentially positive strategy to address issues of climate change and environmental destruction through nonhuman connections, yet both texts present the idea that perhaps we are too late to enact deeper, ecocentric connections with the nonhuman. Even if fiction itself can function as a kind of messenger through the perspectives of its protagonists, the impacts of climate change and environmental exploitation are already being seen. Nevertheless, the positive transformations of the human and the animal through entanglements in both novels suggest a future in which reciprocity and gifting between people and between species is not only desirable, but a necessity. As Kimmerer writes, "the relationship of gratitude and reciprocity … can increase the evolutionary fitness of both plant and animal"; the stories we tell about this kind of respectful reciprocity are essential in developing it (2020, 30).

In the final passages of *The Swan Book*, Oblivia has become one of the ghosts that walk the swamplands of her home, embracing a sense of fluidity that is necessary to an engagement with life in all its forms. *Locust Girl* ends with a final reflection on the coming together of Amedea and the locust, and the power of their shared materiality and the songs they sing together after she escapes the Five Kingdoms: "Can you hear that little flutter? It's an insect heart…. Now you know what we've always shared. No border can deny it" (Bobis 2015, 178). In the midst of this powerful recognition, though, is the realisation that "it's such a long flight across the border" (Bobis 2015, 179). The simultaneously hopeless and hopeful conclusions of these novels therefore indicate a sense of feeling lost—whether or not the border has been crossed or the ideologies that inform its existence are overcome. If no one else follows Oblivia's and Amedea's attentiveness to reciprocity beyond the divide of borders and Western dualisms, then all was for nothing. Perhaps, then, these works encourage the reader to follow these protagonists as they lead by example in the ways that they build connections with the nonhuman, providing a redeeming means by which others living in a time of climate emergency and dominant colonial-capitalist authorities may begin to face such challenges.

References

Ahuja, Neel. 2014. Postcolonial Critique in a Multispecies World. In *Representing the Modern Animal in Culture*, ed. J. Dubino, Z. Rashidian, and A. Smyth, 227–237. New York: Palgrave Macmillan.

Araluen, Evelyn. 2019. To Outlive a Home: Poetics of a Crumbling Domestic. *Cordite Poetry Review*, February 1, http://cordite.org.au/essays/to-outlive-a-home/. Accessed 3 Feb 2025.

Australian Federal Government. 2007. *Northern Territory National Emergency Response Act*. https://www.legislation.gov.au/Details/C2011C00053

Australian Federal Government. 2012. *Stronger Futures in the Northern Territory Act*. https://www.legislation.gov.au/Details/C2012A00100

Barras, Arnaud. 2015. The Law of Storytelling: The Hermeneutics of Relationality in Alexis Wright's *The Swan Book*. *Journal of the Association for the Study of Australian Literature* 15 (3): 1–12.

Belcourt, Billy-Ray. 2020. An Indigenous critique of Critical Animal Studies. In *Colonialism and Animality: Anti-Colonial Perspectives in Critical Animal Studies*, ed. Kelly Struthers Montford and Chloë Taylor, 19–28. London: Routledge.

Birch, Tony. 2015. It's Been, It's Here: Tony Birch on Climate Change's Past and Present. *The Wheeler Centre Notes*, March 23. https://www.wheelercentre.com/news-stories/2015/its-been-its-here-tony-birch-on-climate-changes-past-and-present. Accessed 3 Feb 2025.

Bobis, Merlinda. 2010. The Asian Conspiracy: Deploying Voice/Deploying Story. *Australian Literary Studies* 25 (3): 1–19.

Bobis, Merlinda. 2015. *Locust Girl: A Lovesong*. North Melbourne: Spinifex Press.

Clark, Timothy. 2014. Nature, Post Nature. In *The Cambridge Companion to Literature and the Environment*, ed. Louise Westling, 75–89. New York: Cambridge University Press.

Crosby, Alexandra, Jacquie Lorber-Kasunic, and Ilaria Vanni Accarigi. 2014. Value the Edge: Permaculture as Counterculture in Australia. *M/C Journal* 17 (6): np.

Cudworth, Erika. 2005. *Developing Ecofeminist Theory: The Complexity of Difference*. New York: Palgrave Macmillan.

Griffith Review. 2021. Noel Pearson. https://www.griffithreview.com/contributors/noel-pearson/. Accessed 11 Feb 2025.

Fox, Mem. 1983. *Possum Magic*. Malvern: Omnibus Books.

Gibbs, May. 1918. *Tales of Snugglepot and Cuddlepie*. Sydney: Angus & Robertson.

Gibson, Ross. 1992. *South of the West: Postcolonialism and the Narrative Construction of Australia*. Bloomington: Indiana University Press.

Gifford, Terry. 2008. Recent Critiques of Ecocriticism. *New Formations* 64: 15–24.

Gleeson-White, Jane. 2013. Going Viral: *The Swan Book* by Alexis Wright. *Sydney Review of Books*, August 23, https://sydneyreviewofbooks.com/reviews/going-viral-the-swan-book-by-alexis-wright. Accessed 3 Feb 2025.

Haraway, Donna. 2015. Anthropocene, Capitalocene, Plantationocene, Chthulucene: Making Kin. *Environmental Humanities* 6 (1): 159–165.

Holgate, Ben. 2015. Unsettling Narratives: Re-Evaluating Magical Realism as Postcolonial Discourse Through Alexis Wright's *Carpentaria* and *The Swan Book*. *Journal of Postcolonial Writing* 51 (6): 634–647.

Kimmerer, Robin Wall. 2020 [2013]. *Braiding Sweetgrass: Indigenous Wisdom, Scientific Knowledge and the Teachings of Plants*. London: Penguin Books.

Knowles, Sheena. 1988. *Edward the Emu*. Sydney: Collins Australia.

Leane, Jeanine. 2010. Aboriginal Representation: Conflict or Dialogue in the Academy. *The Australian Journal of Indigenous Education* 39 (Supplement): 32–39.

Leane, Jeanine. 2015. Historyless People. In *Long History, Deep Time: Deepening Histories of Place*, ed. Ann McGrath and Mary Anne Jebb, 151–162. Acton: Australian National University Press.

Lindsay, Norman. 1918. *The Magic Pudding*. Sydney: Angus & Robertson.

Lybeck, Eric. 2017. *The Black Swan: The Impact of the Highly Improbable*. London: Macat International.

Mason, Travis V., Lisa Szabo-Jones, and Elzette Steenkamp. 2013. Introduction to Postcolonial Ecocriticism Among Settler-Colonial Nations. *ARIEL: A Review of International English Literature* 44 (4): 1–11.

McGuinness, Bruce, and Dennis Walker. 1985. The Politics of Aboriginal Literature. In *Aboriginal Writing Today*, ed. Jack Davis and Bob Hodge, 43–54. Canberra: Aboriginal Studies Press.

Mead, Philip. 2016. The Unjusticiable and the Imaginable. *Journal of the Association for the Study of Australian Literature* 16 (2): 1–10.

Mead, Philip. 2018. Unresolved Sovereignty and the Anthropocene Novel: Alexis Wright's *The Swan Book*. *Journal of Australian Studies* 42 (4): 524–538.

Mentz, Steve. 2017. Seep. In *Veer Ecology: A Companion for Environmental Thinking*, ed. Jeffrey Jerome Cohen and Lowell Duckert, 282–296. Minneapolis: University of Minnesota Press.

Merleau-Ponty, Maurice. 1973. *The Visible and the Invisible*. Evanston: Northwestern University.

Mudrooroo. 1997. *Milli Milli Wangka: The Indigenous Literature of Australia*. South Melbourne: Hyland House Publishing.

Murray, Jessica. 2022. Using Critical Animal Studies to Read Climate Change Fiction: Literary Reflections and Provocations. *English Academic Review* 39 (2): 67–78.

Nakata, N. Martin, Victoria Nakata, Sarah Keech, and Reuben Bolt. 2012. Decolonial Goals and Pedagogies for Indigenous Studies. *Education & Society* 1 (1): 120–140.

Nimmo, Dale Graeme, and Kelly K. Miller. 2007. Ecological and Human Dimensions of Management of Feral Horses in Australia: A Review. *Wildlife Research* 34 (5): 408–417.

Ogden, Laura A., Billy Hall, and Kimiko Tanita. 2013. Animals, Plants, People, and Things: A Review of Multispecies Ethnography. *Environment and Society* 4: 5–24.

Perkins, Elizabeth. 1988. Colonial Transformations: Writing and the Dilemma of Colonisation. In *The Penguin New Literary History of Australia*, ed. Laurie Hergenhan, 139–153. Ringwood: Penguin Books Australia.

Potter, Emily. 2017. Teaching Climate Change at the End of Nature: Postcolonial Australia, Indigenous Realism and Alexis Wright's *The Swan Book*. In *Teaching Climate Change in the Humanities*, ed. Stephen Siperstein, Shane Hall, and Stephanie LeMenager, 243–249. London and New York: Routledge.

Potter, Emily. 2019. *Writing Belonging at the millennium: Notes from the Field on Settler-Colonial Place*. Bristol: Intellect Books.
Procter, James. 2009. Reading, Taste and Postcolonial Studies. *Interventions* 11 (2): 180–198.
Rose, Deborah Bird. 2013. Anthropocene Noir. *Arena Journal* 41:206–219.
Sandilands, Catriona. 2017. Vegetate. In *Veer Ecology: A Companion for Environmental Thinking*, ed. Jeffrey Jerome Cohen and Lowell Duckert, 16–29. Minneapolis: University of Minnesota Press.
Simpson, Nardi. 2020. *Song of the Crocodile*. Sydney: Hachette Australia.
Stengers, Isabelle. 2010. Including Nonhumans in Political Theory: Opening the Pandora's Box? In *Political Matter: Technoscience, Democracy, and Public Life*, ed. Bruce Braun and Sarah J. Whatmore, 3–33. Minneapolis: University of Minnesota Press.
Swadener, Beth Blue, and Kagendo Mutua. 2008. Decolonizing Performances: Deconstructing the Global Postcolonial. In *Handbook of Critical and Indigenous Methodologies*, ed. Norman K. Denzin, Yvonna S. Lincoln, and Linda Tuhiwai Smith, 31–43. Los Angeles: Sage.
Takolander, Maria Kaaren. 2016. Theorizing Irony and Trauma in Magical Realism: Junot Diaz's *The Brief Wondrous Life of Oscar Wao* and Alexis Wright's *The Swan Book. A Review of International English Literature* 47 (3): 95–122.
Taylor, Nik, and Richard Twine. 2014. Introduction: Locating the 'critical' in critical animal studies. In *The Rise of Critical Animal Studies: From the Margins to the Centre*, ed. Nik Taylor and Richard Twine, 1–15. London: Routledge.
Van Dooren, Thom, and Deborah Bird Rose. 2016. Lively Ethography: Storying Animist Worlds. *Environmental Humanities* 8 (1): 77–94.
Van Dooren, Thom, Eben Kirksey, and Ursula Münster. 2016. Multispecies Studies: Cultivating Arts of Attentiveness. *Environmental Humanities* 8 (1): 1–23.
Van Neerven, Ellen. 2014. *Heat and Light*. St Lucia: University of Queensland Press.
Van Rijswijk, Honni. 2014. Archiving the Northern Territory Intervention in Law and in the Literary Counter-Imaginary. *The Australian Feminist Law Journal* 40 (1): 117–133.
West, Patrick. 2024. *Architectures of Occupation in the Australian Short Story: Literature and the Built Environment after 1900*. London and New York: Routledge.
White, Jessica. 2014. Fluid Worlds: Reflecting Climate Change in *The Swan Book* and *The Sunlit Zone*. *Southerly* 74 (1): 142–163.
Whyte, Kyle. 2017. Indigenous Climate Change Studies: Indigenizing Futures, Decolonizing the Anthropocene. *English Language Notes* 55 (1–2): 153–162.
Wolfe, Patrick. 1999. *Settler Colonialism and the Transformation of Anthropology*. London and New York: Cassell
Wright, Alexis. 2006. *Carpentaria*. Artarmon: Giramondo Publishing Company.

Wright, Alexis. 2017. *Tracker*. Artarmon: Giramondo Publishing Company.
Wright, Alexis. 2014. *The Swan Book*, B Format Edition. Artarmon: Giramondo Publishing Company.
Wright, Alexis and Arnold Zable. 2013. The Future of Swans. *Overland Literary Journal* 213 (Summer). https://overland.org.au/previous-issues/issue-213/feature-alexis-wright-and-arnold-zable/
Wright, Alexis. 2013. *The Swan Book*. Artarmon: Giramondo Publishing Company.
Wright, Alexis. 2023. *Praiseworthy*. Artarmon: Giramondo Publishing Company.

CHAPTER 4

Collaborative Storytelling and Un/Belonging in the Australian Habitat Story

4.1 "Part of Their Story": Unsettling Habitat

Although the animal is a large part of my discussion of the nonhuman in this book, the nonhuman also encompasses plants, fungi and bacteria, amongst other life forms. The nonhuman may also be a collaborative ecosystem which includes a "habitat"—a natural environment where nonhuman individuals seek shelter, sustenance and connections with fellow nonhumans. Just as Australia is home to a variety of unique native animals found nowhere else on Earth, this continent also exhibits an extraordinary array of native habitats that these animals call home.

It is the exploitation and destruction of some of these habitats that have resulted in the endangerment and extinction of many native animal and plant species. Some habitats in particular are more targeted than others, such as the many hectares of forest which have been, and are still, being logged throughout eastern Australia, primarily for the purposes of agriculture (see Slezak 2018). Additionally, scientists are already observing the ways in which climate change is affecting such habitats on a broader scale. This raises the question of what people choose to value: do we focus on saving one habitat over another and, if so, which should be prioritised? Are we effectively giving up on certain habitats that do not provide a specific kind of value? Val Plumwood's concept of "shadow places" (2008) comes

R. Fetherston, *Theorising the Postcolonial Eco-Novel*, Literatures, Cultures, and the Environment,
https://doi.org/10.1007/978-3-032-04466-2_4

to mind here—a reminder that the way we identify with the natural world can and should encompass those places that are often outside of humans' (particularly settler-colonisers') sphere of self-identification.

William Cronon notes along these lines that

Most of us ... still follow the conventions of the romantic sublime in finding the mountaintop more glorious than the plains, the ancient forest nobler than the grasslands, the mighty canyon more inspiring than the humble marsh. (1996, 22)

Grasslands are a particularly significant Australian example of a habitat that is less "glorious" than others, as they have been highly impacted by colonial agricultural practices.[1] South-eastern Australia's native grasslands remain what Nicholas S.G. Williams and Adrian Marshall describe as a "hard sell" conservation-wise due to their appearance as "drab paddocks" for most of the year (2015, 1–2). Deserts are another largely ignored or otherwise unrealistically represented habitat within Australian settler culture. Libby Robin argues that Europeans coming to Australia had to (and are still having to) come to terms with "what it means to live in a land that is more than two-thirds arid or semi-arid" (2007, 1). Bruce Pascoe aptly describes the "cultural myopia" of settler-colonial pastoralists who believed that, "even as the nature of the country changed, they would never blame their own form of agriculture for that devastation" (2014, 11). Some Australian ecofiction provides a lens through which readers can understand these habitats as not solely dependent on settler-colonial considerations of the land that stem from the Romantic ideal of nature as "spiritualized, always primarily an idea rather than an actual physical entity" (Oerlemans 2004, 33). Therein lies the possibility of pursuing a more complex understanding of how one relates to these habitats, and the humans and nonhumans that live alongside and within them.

This chapter focuses on two novels from within the contemporary Australian settler literature tradition. Specifically, I explore the representation of rainforest habitat in Mireille Juchau's *The World Without Us* (2015) and subtropical coastal habitat in Catherine McKinnon's *Storyland* (2017). *Storyland* in particular addresses issues pertinent to the coastal community of Lake Illawarra and spans centuries, from the colonial past of 1796 to the posthumanist future of 2717. *The World Without Us* depicts the nuances of living amongst mountainous rainforest in a rural New South

[1] For a detailed discussion of the impact of European agricultural practices on native Australian grasslands, see Williams et al. (2015).

Wales town, and how the past influences human interactions with the natural world. Both texts depict the potential future impacts of climate change and environmental exploitation on these habitats, each exploring the passing of time in human and nonhuman communities that are rapidly altered through ecological crises. Whilst *The World Without Us* explores the nature of time, memory and generational trauma in a realist setting, *Storyland* depicts the changing habitat and community of a specific coastal location across hundreds of years, with parts of the narrative being speculative in nature.

It is important to keep in mind that the type of environments depicted in these two texts does not necessarily align with Plumwood's concept of shadow places. Rainforests and some elements of coastal habitat (such as golden, sandy beaches) are, arguably, incredibly well-documented and aesthetically appealing habitats, in a classically Romantic—and colonial—sense. However, the ways in which these habitats are depicted—their presence on the fringes of settler societies, rather than as deep wilderness—suggest that in some ways, such places are still seen as shadows of non-Indigenous Australian communities. Paul Carter writes of coastal zones or, more particularly, "terraqueous zones" (2009, 174) such as those incorporating swampland, as difficult habitats to understand. He states:

It is easy to narrate the life of a river. Flowing rivers share the desire of storytellers to get somewhere. But swamps flow nowhere. Nor, on the other hand, do they provide a solid foundation, a theatrical setting for decisive events. Not exhibiting direction, they resist linear logic. Offering no firm footing, they undermine reason. (2009, 174)

Embracing rather than ignoring these places represents a transformative shift towards a multifaceted ecological awareness that embraces the fluid and diverse nonhuman identities encapsulated in Australian habitats, rather than assuming the position of the traditional settler-coloniser who saw these habitats as conquerable.

As this chapter focuses on the concept of "habitat" as a representation of the native Australian nonhuman, it is worth clarifying how I am utilising this term. Understanding habitat is tied to how the concept of wilderness has been developed, and I would like to make clear that when I discuss Australian habitats, I am considering them as both affecting and affected by Australia's built environment, and not as a separate, hypersacred wilderness existing in contrast to Australian modernity. I speak of a depiction of habitat as something that tells both human and nonhuman stories, and is not merely a symbolic lens through which settler-colonisers might better understand what Cronon describes as "our own unexamined longings and desires" (1996, 7).

I refer to these depictions of habitat as "habitat stories": collaborative stories told through both nonhuman and human experiences that facilitate more nuanced understandings of the relationship between them, creating a space in which habitat itself as a nonhuman entity may exercise agency. It therefore relates to, and is influenced by, certain Indigenous epistemologies, in particular the Indigenous Australian concept of Country as articulated by Marcia Langton, Bruce Pascoe, and Aunty Joan Tranter, amongst many others. One particularly relevant aspect of Country is that described by the Gay'wu Group of Women in the book *Song Spirals*:

We don't have an identity without Country. We are all connected. Every contour on the land, every rock, every water, is connected to us. They are bigger than us, we are just a little thing. (Burarrwana et al. 2019, 149)

I am not putting forward this idea of habitat stories in a way that co-opts Indigenous ideas of Country, but rather as a concept that is influenced by and reflects some aspects of it. I believe the term "habitat stories" is useful from a literary studies perspective in achieving a deeper understanding of how habitat is critically deployed in Australian literature, particularly Australian settler-colonial literature, as it has so far been a largely overlooked aspect of such texts, even in fiction that explores explicitly environmental themes.

Habitat and the wilderness have been seen in the past—and by many still—as a stage for colonial domination; a place to either exploit for resources or retain for (a particular kind of) natural beauty. Langton presents a similar argument to Cronon in an Australian context, linking the depictions of Indigenous Australian people in settler-colonial culture with ideas of colonial dominance:

Representations of natives and landscape, especially those which are symbols of conquest and colonial history, are bound up in Australian artistic and cultural traditions, especially as themes of pioneer dominance and progress. (1996, 12)

The way that I think about the depiction of habitat in Australian ecofiction is linked to this, as *Storyland* and *The World Without Us* both, to varying extents, endorse stereotypes regarding the settler-colonial symbology that is forced onto the land. But I also expand beyond this in my reasoning that Australian settler authors writing in the past decade, such as Juchau and McKinnon, utilise representations of habitat to engage with issues of colonisation and environmental degradation. They do not use such habitats as mere metaphor for the human, or colonialist, condition, but critically deploy habitat stories in an effort to shift contemporary understandings of settler-colonial belonging. Such stories unsettle notions of settler belonging as traditionally linked to places bearing the mark of Western-colonial domination, such as Euro-agricultural landscapes, fenced property, paved roads and built-up towns and cities. As Langton argues, "in manufacturing representations of 'Nature' and 'wilderness', the human mind constructs a relationship with the nonhuman" (1996, 19). How, then, do settler-colonial texts like Juchau's and McKinnon's represent this relationship? Do they challenge the colonial idea of the Australian wilderness, thereby altering the way settler-colonisers may perceive the nonhuman, or do they adhere to Langton's concept of manufactured representations?

To understand the depiction of the Australian nonhuman in settler literature, we must look to the authors' portrayals of habitat as a means through which to understand the European colonial impact on these places but also how the settler does or does not exert belonging in the context of the modern environmental crisis. There is a need to better connect non-Indigenous Australians with the notion of land as a human *and* nonhuman habitat (an important aspect of Indigenous notions of Country, as defined by Aunty Joan Tranter (2013) and the Gay'wu Group of Women (Burarrwana et al. 2019), rather than simply an exclusively *human* realm. This is one reason I have chosen the term "habitat" over "place" in this chapter, as I believe more attention must be paid to how settler Australians consider their influence on habitats, as well as the influence that habitats have on them in terms of the resources they provide and the cultural significance they may denote.

Such a perspective is inspired by Robin's claim (which she has drawn from the writings of Aldo Leopold) that habitats—such as mountains and deserts—are "nonhuman actors" within Australian history and have played a significant role in Australia's evolving national identity (2007, 2). These actors play a role in Australian settler stories, too, by interrogating the connections between non-Indigenous Australians and habitat. Through this understanding of habitats as not just places in which stories are told but as places with stories in their own right, habitat stories expand the current literary dialogue on place in Australian settler-colonial literature to include a deeper consideration of the role that habitat plays as a character itself, and not simply as a setting.

The habitat stories in this chapter's texts can and do exist separately to those human stories that may take place within habitat; habitat stories take place before, during and after practices of habitation. Each story is also multi-vocal, not narrated by one but by many living beings, sometimes including a human perspective but often told through the habitat itself and from the perspective of an individual nonhuman or many nonhumans within this habitat. The work of contemporary settler-colonial authors whose writing includes such considerations marks a move away from more traditional portrayals of the nonhuman in Australian settler literature.

Australian settler literature is concerned with what Jeanine Leane describes as "journeys into the interior of the settler mind and consciousness and its understanding of the phenomena of 'the Aborigine' and … questions of authority and power" (2014, 1). As ecofiction, *Storyland* and *The World Without Us* directly and indirectly engage with this history of settler-colonial literature through portrayals of the human (and primarily non-Indigenous) connection with the natural world. More specifically, they address which group—typically, the colonisers—exhibits authority over Australia's native habitats, and at what cost such authority is exercised. Leane suggests that "the British might be 'settled' but the journey for settlement and resettlement is ongoing" (2014, 2). Some settler literature therefore reflects what Leane describes as a "quest to belong" and a "changing consciousness of Aboriginal presence" (2014, 2). As works of Australian settler literature, both novels in this chapter advance this notion to some extent. However, this is achieved through a critical perspective of the ways in which colonial domination of nature may be overcome via a deeper sense of non-Indigenous belonging within native Australian habitats.

Both novels consider the idea that "the ecocidal logics that now govern our world are not inevitable or 'human nature'" but are rather the consequence of colonisation (Davis and Todd 2017, 763). In some ways, the texts analysed in this chapter challenge the notion that the destruction of the natural world is inevitable, given both authors' emphasis on understanding and learning from past mistakes. This suggests that the climate crisis is not unavoidable but is changeable if humans or, in this context, non-Indigenous Australians, seek deeper connections with the nonhumans that European colonisation have decimated or altered in the development of extractivist industries. Given that both Juchau and McKinnon are non-Indigenous Australian authors, I analyse their texts with this relationship between colonisation and the climate crisis in mind.

Storyland and *The World Without Us* are both complicit in and challenge the common traits of settler literature. Although reflecting non-Indigenous attempts to belong on Indigenous land (arguably the primary premise of most if not all Australian settler literature), these texts also challenge the coloniser's problematic relationship with the Australian nonhuman, offering alternative worldviews that delve deeper into a more ecocentric relationship with the natural world. Emily Potter is critical of the ways in which Australian settler literature provides a service of forging naturalised, assured relations between non-Indigenous Australians and place. However, she contends that there is a tradition within Australian literature—one that is still evolving—that seeks less certain engagements with the land that settlers inhabit. Deeper understandings of these "expression[s] of disconcerted non-indigenous belonging" within Australian settler texts may allow for a "dethroning" of the power held by white, non-Indigenous Australians to "maintain and reinforce silences and exclusions" (Potter 2019, 9). Both *Storyland* and *The World Without Us* sit within this tradition as they advance the idea of an unsettled sense of belonging amongst settler Australians, particularly in the context of native Australian habitats.

As may already be implicit in my definition of habitat, it is important to acknowledge that the native nonhuman is not something restricted to the animal, or to individual species. The nonhuman is arguably always collaborative, each organism relying on others within distinct but also linked ecosystems. The Indigenous Australian concept of Country is of relevance here, as it was in Chap. 3, but for somewhat different reasons. Country represents a way of thinking about the natural world and the human relationship with it in a multifaceted way. In an interview with Anne Brewster, Noongar author Kim Scott remarks on one of his fictional characters who has great confidence in himself and his heritage, stating that "he just does not have it in his mind that anyone could ever want to conquer another's country, because he was so connected with it: you're the same, you and your country" (2015, 4). There is perhaps room for Australian settler authors to broaden their consideration of what "conquering" means in this context. For example, in what ways is the search for non-Indigenous belonging on Indigenous land a form of dominance, and can less dualistic understandings of habitat (as something that is not considered apart from the human) assist with such considerations?

Ensuring the survival of Country is not simply a means of ensuring the survival of the human self, but also the "spiritual, cultural and economic survival" (Brewster 2015, 4) of the lands themselves. These lands possess cultures and stories of their own, and the way that these stories are told differs vastly between Indigenous Australian peoples and settler-colonisers. This is poignantly expressed by Ambelin Kwaymullina in her manifesto, *Living on Stolen Land* (2020), written in verse:

> For example
> ask a Settler reductionist thinker
> to explain a hill
> and they might speak
> about its height
> about the composition
> of the rock that form it
> about the first Settler
> to reach the peak …
>
> Ask an Indigenous thinker
> to explain a hill
> they might speak
> about the river that runs past it
> the plants and animals
> to whom it is home
> about the earth below
> the turning stars above
> about the Ancestor
> who is the hill
> the Ancestor hill
> who is family (22–23)

This Indigenous thinking is strongly reflected in Wright's *The Swan Book* (2013) in Chap. 3, Heiss' *Bila Yarrudhanggalangdhuray* (2021) in Chap. 7, as well as in the fiction of Kim Scott, Tara June Winch, and many other First Nations authors. Such Indigenous epistemologies as read in these texts may have influenced the direction of non-Indigenous authors such as McKinnon who incorporate less Euro-Western understandings of Australian habitats in their writing. McKinnon explicitly identifies this influence in the acknowledgements of *Storyland*, recognising "the *Elouera* and *Wadi Wadi* (also *Wodi Wodi*) people of *Dharawal* Country" (2017, 379) (where her novel is set) and also thanking Aunty Barbara Nicholson, a Wodi Wodi Elder, for "local knowledge" (380). While Juchau does not incorporate Indigenous understandings to the same extent as McKinnon, she has recognised the existence of "competing voices, competing stories" in understanding the climate crisis (Hardy 2015), which may account for her critique of the homogenising settler-colonial stories of place and habitat in *The World Without Us*.

In the past, Australian settler literature has focused on presenting Australian habitats as antagonists. In a similar vein to how Langton describes tropes of Australian artworks which depict the nation's colonial history, the protagonists of Australian settler literature are often framed as "heroes and heroines who face the adversary" (1996, 11). Degrading images of the declining populations of Indigenous Australian people, as shown in colonialist artworks and depicted in what Andrew McCann refers to as "the literature of extinction" (2006), were at one point replaced by depictions of "heroic pioneers and their stoic wives, their adversary now a harsh landscape, tamed by axes and bullock teams" (Langton 1996, 13). Langton argues that this colonialist imaginary demonstrates a reluctance to depict Australia's landscapes in anything other than Eurocentric terms (1996, 13). Just as such landscapes have been the focus of Australian settler art since the early nineteenth century, similar tropes of depicting Australian habitats as places to dominate and control are seen in Australian settler literature. Langton goes on to describe how these same habitats and their nonhuman inhabitants, as well as Indigenous Australian art and design, were then celebrated and became characteristic of Australian post-war modernism, featuring on the nation's coins and stamps and "creat[ing] a mythology which allows its participants to presume innocence in the colonial process" (1996, 16). This effectively began the incorporation of images such as the kangaroo and the platypus into designs and imagery that promoted a post-war Australian way of life, representing what Langton defines as "an imperial signification of the dominion of civilisation over nature" (1996, 16).

The tradition of writing more complex human–nonhuman relations into Australian settler literature using the concept of habitat is therefore relatively recent. In their book *Imagined Landscapes* (2015), Jane Stadler et al. discuss how the qualitative information provided in Australian narratives of the past, namely dramatised historical accounts, "can offer insight into evolving perspectives on a locale over time" (2015, 97). This idea is particularly pertinent in the context of *Storyland*, as McKinnon presents a fictionalised account of George Bass, Matthew Flinders and Will Martin's second journey on the *Tom Thumb* along Australia's east coast in 1796. It is more broadly relevant too, as both texts in this chapter, whilst fictional, depict the very real changes that have occurred and are occurring within two distinct native Australian habitats, making the information they convey about environmental crisis and the plight of the nonhuman a potentially reliable indicator of how non-Indigenous Australians view two specific Australian locales.

How, then, can we define what might be termed "narrative landscapes" within the contemporary Australian settler literature tradition? Could these be the places in which stories happen or do the stories come from habitats? Using the term "habitat" in place of "landscape" implies the latter, suggesting that humans are both within and outside of these places and are not simply living *on* them. Returning to the above passage from *Living on Stolen Land*, settler-colonial thinking about habitat can be classified as "reductionist" (Kwaymullina 2020, 22) in its perception of each natural place or nonhuman existing in a vacuum and as solely symbolic of settler culture and achievements. Indigenous thinking instead engages with a broader interconnectedness between "the hill" that Kwaymullina speaks of, and the myriad other beings that exist alongside and within it to form a "family" (2020, 23). This latter perspective is the lens through which I would like to understand the way that habitat is evoked in McKinnon and Juchau's novels.

4.2 The Coastal and the Estuarine in Catherine McKinnon's *Storyland*

Shortlisted for the 2018 Miles Franklin Award, *Storyland* is McKinnon's second novel and reflects her interest in telling what Linda Morris describes as "the untold perspectives of history" through a series of "five interlocking stories spanning a millennium" (2017). Morris also remarks on the cohesion across the stories as represented by the natural surroundings:

Common to these diverse stories is Mullet Creek, the brackish Lake Illawarra and the waterway Flinders named Canoe Rivulet which feeds into it, located two hours south of Sydney in the lee of the former coal town of Wollongong. (2017)

The human and nonhuman come together in surprising ways in this novel, through both habitat and memory. Each chapter of *Storyland* is set in a particular time period in Australian history and revolves around a significant event in the life of each chapter's protagonist. Reflecting the narrative strategy employed by British author David Mitchell in his speculative novel, *Cloud Atlas* (2004), McKinnon's text moves between various stories set during different periods of Lake Illawarra's history. Beginning with the story of Will Martin—a fictional portrayal of the real-life servant to British explorer George Bass—and ending with the story of Nada, the narrative of *Storyland* remains in the same geological location of Lake Illawarra, but spans the years and cultures of Australia's past and potential future. McKinnon's implementation of a similar narrative structure to Mitchell's work is effective in linking colonisation, Indigenous dispossession and the climate crisis. The narrative both begins and ends with a story of the early years of Australia's colonisation, and the stories in between highlight the ensuing environmental devastation and colonial oppression of Indigenous people, culminating in a dystopian future where climate change has irrevocably changed this region of Australia. Perhaps more significantly, though, the structure draws attention to the possibility of avoiding such an apocalyptic future. As Diletta de Cristofaro says of *Cloud Atlas*, this "concertina-like structure" challenges the "apocalyptic determinism" of history, particularly as told by the West (2018, 243), and "gives prominence to the individual's agency to shape the future" (248). McKinnon therefore positions the individual as capable of challenging the dominant systems of colonialism and seeking out a better future than the one that colonisation has so far determined in Australia and beyond.

The novel begins with a fictionalised account of the journey of the *Tom Thumb* from the perspective of Will Martin. Set in 1796, this initial and later connecting chapter[2] of McKinnon's text addresses the issue at the heart of much of this novel: the impact of colonisers on Indigenous Australian cultures and the Australian nonhuman. Will reflects on the encounters between Bass, Matthew Flinders and the Indigenous Australians they meet along the coast. The story that follows is set in 1822 and demonstrates the violence enacted against Indigenous Australians by early British settlers, told from the perspective of a convict named Hawker. The third story in the novel depicts Lola, a teenage girl who, in 1900, searches for a missing friend whose body she later discovers in the waters of Lake Illawarra. Lola is an Indigenous Australian, and her story again depicts the

[2] As in, each story in McKinnon's text is separated into two chapters—one in the first half of the novel which introduces the story and the second in the latter half of the novel which concludes the story.

violent, domineering crimes of non-Indigenous Australians committed against Indigenous people and the surrounding environment. The next story, set in 1998, presents a more diverse community of non-Indigenous Australians living near Lake Illawarra from the perspective of Bel, a young girl who spends time exploring the natural surroundings with her friends. The final story is set in both 2033 and 2717 and follows Nada, a future human attempting to survive the calamitous impacts of climate change on her community and her retelling of these events to those collecting stories from the past in 2717. The steady presence of an immense native fig tree permeates every chapter of *Storyland*, each human generation forming a different relationship with this tree and the surrounding land and waterways of Lake Illawarra.

The focus habitat of *Storyland* is Lake Illawarra, a coastal estuary located 80 kilometres from Sydney which features a coastal barrier that divides the area's estuarine water from the Tasman Sea (Deng et al. 2018). It is therefore a habitat that may be particularly threatened by rising seawaters as a result of climate change—an event which is explicitly explored in McKinnon's novel. In a similar way that the rainforest habitat of Juchau's novel exists alongside a human community, so too does Lake Illawarra exist alongside the New South Wales city of Wollongong.

As indicated by the novel's title, a large focus of McKinnon's text is the importance of story and how it influences human understandings of the natural world across time. As referenced in this book's introduction, the novel is prefaced with a quote by Mark Rowlands from *The Philosopher and the Wolf*:

If I wanted a one-sentence definition of human beings, this would do: humans are the animals that believe the stories they tell about themselves. Humans are credulous animals. (2009, 2, cited in McKinnon 2017, epigraph)

The relevance of this quote to McKinnon's exploration of the human–nonhuman relationship is significant. The very human stories in *Storyland* are centred around a very nonhuman entity: the small-leaved fig, or *Ficus obliqua* (McKinnon 2017, 176), a native Australian tree which, as revealed in Nada's future story, is "a thousand years old" (183–184), signifying that "perhaps this place has always been safe" (184). The fig tree stands at the centre of the various lives portrayed in *Storyland*, even if this is not explicitly recognised by the characters themselves. It is this tree which prompts the memory of Nada as a regenerated human in the year 2717 (McKinnon 2017, 176), and is the tree which is the site of curiosity and adventure for Bel and her friends (for example, Bel consistently admires the historical and natural complexity of the area, if even from a young, somewhat naïve perspective—such as how "the lake was once full of fish and it was a refrigerator for everyone who camped on the banks in the olden Aborigine days" (McKinnon 2017, 123)). Although an individual nonhuman, the fig tree represents the broader habitat story embedded in this narrative, encouraging the reader to question what this tree and the habitat in which it resides experience as the colonisation of Australia progresses over time, culminating in the devastating effects of climate change. In this sense, the Lake Illawarra habitat is a character that acts and reacts alongside the human characters of the novel; it is one of many "nonhuman actors", as Robin would describe (2007, 2), that play a role within settler-colonial history.

Such an approach is more in line with Indigenous epistemologies (for instance, the work of Tranter (2013), Pascoe (2014) and the knowledge written about by non-Indigenous scholar Deborah Bird Rose following her interviews with Indigenous Australians (2013)), emphasising the material nature and experiences of the tree and its habitat (what are the stories occurring in and around this nonhuman place?) rather than the metaphoric nature of the tree in relation to the human elements of the story. McKinnon's fig tree and the surrounds of Lake Illawarra largely function as characters themselves within the narrative, and are not simply symbolic of what has been projected onto them by the human imaginary.

Notably, habitats and their nonhuman inhabitants are crucially linked to human stories throughout McKinnon's novel, as demonstrated by one character's remark that "We're part of their story, not the other way around" (McKinnon 2017, 247). Indigenous writings about Country suggest something similar, as indicated by Tranter's comments that Indigenous Australians must ensure the survival of the land in "spiritual, cultural and economic" ways (2013, 11), without which people themselves would not survive. This idea is further emphasised by the role that the nonhuman often plays in the transition from generation to generation, story to story. The conclusion of each chapter is cut off mid-sentence, allowing for a flow-on effect into the following chapter which begins part-way through a different, but potentially related, sentence. From British convict Hawker's story in 1822 to Indigenous teenager Lola's story in 1900, the transition is represented by the Australian boobook owl: the final words of Hawker's story being "An owl", and the first words of Lola's story being "*boo-book*" (McKinnon 2017, 68; italics in original).

The reader experiences a similar transition between the second part of Hawker's story—"The flap of wings above, pelicans flying"—and the second part of Will Martin's story—"across the sky": read completely as "The flap of wings above, pelicans flying … across the sky" (McKinnon 2017, 315).

These nonhuman transitions suggest that the interconnections felt between humans across generations are not just facilitated by their nonhuman surroundings, but perhaps instigated by them, speaking to the stability of nature—described by Nada when she states that the resilient fig tree made her home feel "safe" (McKinnon 2017, 184)—but also the unpredictable behaviour of humans living in this habitat, each story demonstrating the capacity for human violence and environmental destruction during times of colonisation. A striking example of this occurs in Lola's story, when the body of teenager Jewell is finally found:

We wade to where the reeds are higher than our heads, clearing away fallen branches. I pull back the thick reeds, and see her. Jewell. Face down in the water. She is tied to an old rotten boat that has sunk. (McKinnon 2017, 280)

Jewell is found in Mullet Creek, a waterway connected to Lake Illawarra and surrounded by dense bush. On finding her body, this violent act is now inextricably connected to this habitat but it is significant that she is "tied to an old rotten boat" (McKinnon 2017, 280)—a clear sign of settler-colonial culture in the Australian bush. This distinguishes McKinnon's novel from the settler-colonial trope of portraying the natural world itself as the perpetrator of violence, instead demonstrating the strong link between colonisation and increased tension, assault and murder.

Whilst the settler-colonial human may be responsible for violent acts against fellow humans and the nonhuman, the nonhuman habitat itself presents an agency that goes beyond its role as a mere backdrop to human events. The nonhuman is not simply a presence in this novel but is depicted as interrogating complex human–nature relationships. In Nada's story, this is made particularly clear when one character recalls his Aunty's thoughts on the white-bellied sea eagle:

When I was young my very old Aunty told me that the white-bellied sea eagle knew about the meeting of water and land. She said, you must *learn* from them boy. (McKinnon 2017, 228; italics in original)

In this scenario, the nonhuman is granted agency in their ability to educate humans about the land. Much like Robin's claim, this is a "nonhuman actor" like the fig tree, guiding the trajectory of certain events in

Australia's history (2007, 2)—both those of the past, and those yet to come. Understanding these human–nonhuman connections means reading the land, the habitat. The tagline of McKinnon's novel aptly represents this: "the land is a book, waiting to be read" (2017).

McKinnon often draws direct parallels between human history and habitat throughout the novel, at one point describing how "human events are like rivers, they change course depending on circumstance" (McKinnon 2017, 314) and, most poignantly, in the final words of the novel that conclude the story of Will Martin in the year 1796: "Now I am a man, and a man is his story told. I say this out loud, but only the nightjar answers" (369). The nightjar, a bird that would likely still find shelter in the habitat of Lake Illawarra today, is the only respondent to Will's own personal history, and the broader history of colonisation in Australia at this moment in time. The nightjar stands both within and outside of Will's story, responding to his experiences but its call also signifying that it has been here all along, both before and during Australia's colonisation.

The idea that McKinnon's narrative is firmly rooted in habitat is reinforced to the reader by the inclusion of two maps following the novel's final chapter. The first map shows the route of Will's journey in 1796, and the second shows the locations that Bel and her friends visit in 1998. All of the locations as depicted on these maps represent habitats within a larger habitat, as well as ecosystems within ecosystems, each supporting a variety of life—both human and nonhuman—across time. The inclusion of the maps post-narrative rather than in the preface is significant, suggesting that to make sense of the maps in both their historical and fictional contexts, the reader must first experience the stories of the habitats and the connections across time that can be drawn by understanding how various humans and nonhumans have inhabited and made sense of what it means to live in this region.

However, it should be noted that the fig tree is still deployed by McKinnon as a representational device; what Philip Armstrong would describe as a reduction of the nonhuman, in this case a tree, "to a blank screen for the projection of human meaning" (2008, 3, cited in Donovan 2016, 95). In this sense, the fig tree is a projection of the symbolism that settler-colonisers have assigned it over the years as a resilient nonhuman, standing strong in the face of adversity. Within more traditional settler-colonial literature, the inclusion of a nonhuman such as a tree is often representational; a symbol of endurance in place that reflects the characters and events of the narrative as they live in and occur around it.

McKinnon is therefore complicit in continuing this element of settler-colonial storytelling practices, but is also indicating a shift in contemporary settler literature as a literature that may be capable of telling habitat stories in more nuanced ways.

4.3 Temporalities of the Habitat Story

Storyland entwines the idea of a complex, agential habitat with the power of human collectivism in times of crisis. This reveals a focus that Stephanie LeMenager argues many texts within the cli-fi genre (which some might classify Juchau's and McKinnon's as belonging to) lack:

The need for cli-fi is profoundly social, although its realization rarely includes a vibrant evocation of collective, rather than individual experience. (2017, 223)

McKinnon's work engages with this notion of the collective experience of confronting the impacts of climate change. This engagement goes beyond the stereotype of a lone settler surviving the harsh reality of the Australian wilderness, and instead portrays the collective struggle of both humans and nonhumans across generations to adapt to the developing climate emergency within the frame of colonisation.

The narrative of *Storyland* jumps from generation to generation, century to century, providing a complex and strongly connected series of stories that relay the impact of settler Australians across time, in one place. Like Mitchell's *Cloud Atlas*, McKinnon returns to each earlier story in the latter half of the novel—for example, the first half of Will Martin's story is introduced in the novel's first chapter, and the second half is then told in the final chapter of the novel. The same formula is applied to every other story in the text, ensuring that the first half of the novel moves forward into the future whilst the second half moves backwards into the past. In her work on Indigenous Australian concepts of time, Potter describes how, in Indigenous worldviews, time can be defined by "environmental happenings that signify time's enactment" (2015, 102). McKinnon's narrative structure draws on this idea because the novel's stories are not simply classified by the characters that narrate them; they are strongly tied to the different depictions of Lake Illawarra and the changes this area experiences over time. Each chapter is defined by the stage at which environmental degradation, on the part of settler-colonisers, is taking place in that particular time period. It is also significant that so many of the shifts in time from chapter to chapter are characterised by nonhuman sights and sounds in the Lake Illawarra habitat, as described above, suggesting that time extends beyond the linear experiences of some humans to encapsulate how the habitat and its inhabitants both change and remain constant throughout the narrative's chronology.

Nada's story is the only exception to the narrative's chronological structure, spanning the middle section of the novel as an entire, rather than halved, chapter. Her narrative takes place in both the years 2033, when she is alive, and 2717, when she is given a "second life" (McKinnon 2017, 238) by a technologically advanced, future human society. This forms the climactic centrepiece of McKinnon's novel; it is an end point which all of the previous histories have reached, representing a possible outcome for modern Australian society should national action on climate change remain non-existent. The fact that the "Storyland Project" (McKinnon 2017, 174) is referenced in Nada's narrative also hints at this particular story's centrality in the wider context of the novel. However, it also represents a new beginning; a chance to move back in time—as the latter half of the book does—to remedy the ills which have led Australian settler society to the point of climate catastrophe.

McKinnon's use of a non-traditional narrative chronology shows a deviation from the traditional settler narrative in which belonging is sought via the linear domination of habitat; a story with a beginning (feeling lost in an unknown wilderness), middle (domination of nature) and end (achievement of belonging). Instead, *Storyland*'s chronology allows for a more complex understanding of non-Indigenous connection to the nonhuman, affording a nonlinear means of working through the violence and environmental exploitation enacted by colonisers. It provides a more flexible space in which readers are confronted with the crimes of colonisation and simultaneously reassured that we are yet to reach 2033, the year of climate catastrophe in Nada's story, but that this time is not far off from the present. McKinnon suggests that the solution lies in Australia's past; by understanding the events and traditions that have led to this point, modern Australia may be able to bring non-Indigenous thinking about the nonhuman to a more ecocentric place. This nonlinearity of the novel's chronology somewhat reflects aspects of the temporality of the Dreaming, which Potter describes as "dynamic and nonlinear" (2015, 103). Similarly, Billy Griffiths refers to the Dreaming as "a concept so subtle and pervasive that it eludes Western understandings of 'time' [and] 'history'" (2018, 6). However, it is important to note that the structure is also reminiscent of *Cloud Atlas* and other contemporary post-apocalyptic narratives, such as

Jeanette Winterson's *The Stone Gods* (2007) and Emily St John Mandel's *Station Eleven* (2014), which "articulate temporalities critical of Western modernity's apocalyptic understanding of history" (de Cristofaro 2018, 243).[3] I therefore hesitate to entirely apply understandings from the Dreaming to McKinnon's narrative structure, as the nonlinearity of her novel could more be a reflection of how some non-Indigenous, Western authors (of which McKinnon is one) have been writing post-apocalyptic narratives over the past two decades.

As the earliest and latest of all of the novel's stories, it is important to consider how Will's and Nada's stories intersect. Are they both ignorant settler-colonisers, or is Nada an example of a future Australian who is now more in touch with the role her ancestors played in Australia's ecological devastation? The role of Nada's story is to potentially show settler-colonisers the apocalypse that Indigenous peoples have already experienced, while Will's story represents the beginning of this dystopia for Indigenous Australians. Kyle Powys Whyte explains that "Some Indigenous peoples … offer the idea that we confront climate change having already passed through environmental and climate crises arising from the impacts of colonialism" (2018, 226). Nada's conversation with her partner, Ben, at the conclusion to her story reflects how Indigenous people would possibly have felt at the onset of colonisation: "All our family and friends, will we ever see them again? … Will we ever meet strangers again and not be wary of them? … from now on everything is different" (McKinnon 2017, 236). In this sense, *Storyland* comes full circle, beginning and ending with the story of Will in the early years of British colonisation as Indigenous Australians first experience the violent and confronting changes forced upon them, and ending chronologically with Nada's story of environmental apocalypse as she watches the world she knows changing around her, with little or nothing she can do about it.

The cyclical nature of this narrative structure is significant, suggesting the potential for history to repeat itself (as seen across multiple chapters in the novel that detail violence against Indigenous people) and the differences in how Australians have interacted with the nonhuman across generations. This reflects what Potter describes as the Indigenous concept of time as being understood through ecological and ancestral connections (2015, 102). If we examine McKinnon's portrayal of Indigenous

[3] De Cristofaro also discusses other post-apocalyptic texts like *The Stone Gods* and *Station Eleven* in her book *The Contemporary Post-Apocalyptic Novel* (2020).

characters, we see their relationship with the natural world changing over the span of centuries due to the impact of colonisation. The Indigenous people that Will Martin, George Bass and Matthew Flinders encounter are depicted as living in the bush, largely separated from the European settlements of early colonised Australia, whilst later depictions show Lola and her family as living amongst a community of white Australians. The violence enacted against Indigenous Australians escalates as the novel progresses, suggesting that as the colonisers' impact spreads and the surrounding habitats are changed, so too does the destruction of Indigenous life and culture intensify.

While the nonlinear chronology of *Storyland* may represent a break from more traditional Australian settler narratives, aspects of Western time still play a role. Why, for example, does McKinnon choose to begin with the narrative of Bass and Flinders? Apart from perhaps being a method by which to equate the beginnings of Australia's environmental degradation with the beginnings of colonisation, it also hints at a "year zero" (Potter 2014, 107). Where are the pre-colonial stories of Australia in this narrative? Although including the perspective of Indigenous teenager Lola, her experiences still take place within a colonised landscape, and while the novel's chronology is structured in a way that emphasises and critiques the violent history of Australia's colonisation, it is perhaps problematic in its suggestion that colonisers will inevitably dominate both Indigenous people and the land throughout time, as there are no stories that take place outside of this colonisation. Additionally, it is significant that most if not all of the violence in McKinnon's text occurs in the second half of each character's story, rather than the initial chapter in which the reader is introduced to them. This builds tension throughout the novel, but also indicates another inevitability that as time passes, the likelihood of violence increases, indicating what may be construed as a kind of settler determinism: the idea that settler-colonial belonging in Australian place is naturalised and inviolable, and similar to what Potter describes as a "spatial determinism" within settler-colonial practice that enables "cultural forgetting" rather than "an experience of living in history" that recognises colonial violence (2019, 104).

4.4 Settler Representations of Indigenous Experience on Country

An important aspect of McKinnon's novel is the representation of Indigenous Australian experience and its significance within the text's broader narrative of climate change and ecological crisis. It is established early in the narrative that the colonial impact on the New South Wales coast is dramatic, even as early as 1796. Will describes the Tank Stream as "once ... a pure stream, but now ... fouled by pig dung, food scraps and more, so that our small colony is desperate for clean water" (McKinnon 2017, 2). Comparatively, the areas outside of the colony where Will later encounters a local Indigenous group are portrayed as supposedly untouched, describing how "the white sandy beach is curved like a butcher's knife [and] [s]crubby trees beyond become a green blanket covering the hills. A man might get lost in such a place" (McKinnon 2017, 18).

Three of the five story arcs of *Storyland* include Indigenous characters. However, it must be noted that their inclusion also serves a settler-colonial purpose. With the exception of Lola's story in 1900, the other Indigenous characters are not first-person narrators but rather play the role of more minor characters within the narratives told by settler narrators. This is emblematic of the position this text holds within the Australian settler literature tradition, reflecting the norm of works such as Jeannie Gunn's *We of the Never Never* (1908) and James Vance Marshall's *The Children* (1959) (later titled *Walkabout*) that foreground the settler-colonial experience but exploit the Indigenous experience as a means through which non-Indigenous characters reflect on themselves and their relationship with the land.

Leane makes the following statement regarding Kate Grenville's *The Secret River* (2005), which could be accurately applied to the second story of McKinnon's novel:

> ...it writes an alternative settler myth for foundations of emancipists. In this way it is an ongoing investment in nationhood and the reshaping of settler identity in the twenty-first century. (2013, 13)

In a similar vein to William and Sal Thornhill in Grenville's novel, the character of Hawker in *Storyland* demonstrates the hardships of attempting to make a living within the Australian bush, and the tense relationships between the colonisers and Indigenous people. Like this section of McKinnon's text emphasises the increasing violence enacted against Indigenous people—in particular, women—by European colonisers, and the contempt shown towards those Europeans who forged more positive relationships with them, as shown by Kent who is described as "too friendly with the natives by half" (McKinnon 2017, 53). It is significant that Hawker, who exhibits extreme violence towards the local Indigenous people, views his hut as "protection from all that surrounded us, not just natives, the forest too. For the forest is thick with giant trees and ferns and vines, and also with unnatural animals and birds" (McKinnon 2017, 55). Hawker's refusal to positively engage with Indigenous people and his fear of the surrounding habitat solidify his own warped sense of belonging as a colonial frontiersman, just like the many other "heroes and heroines who face the adversary" (Langton 1996, 11) of the Australian bush and its original custodians. Hawker's story can be read as a condemnation of this model of settler-colonial belonging, but it is still problematic in the same sense that *The Secret River* is: it tells stories of Indigenous dispossession and genocide through the eyes of a coloniser.

There is also the issue of whether Indigenous Australian people are present in McKinnon's portrayed future. Are Nada or Bel descended from past Indigenous characters, or does the multicultural nature of these future Australian societies take the place of any meaningful portrayal of Indigenous people? This is another aspect of McKinnon's novel that fits a stereotype of Australian settler literature, as its focus largely remains on the experiences of non-Indigenous people finding belonging on Indigenous land. *Storyland* is then perhaps still an example of the settler literature that Leane describes: "journeys into the interior of the settler mind and consciousness and its understanding of the phenomena of 'the Aborigine'" (2013, 1). Hawker's chapters emphasise this also; his voyeurism towards Indigenous women, his violence and his inability to connect with the surrounding habitat are all symptomatic of a text that falls within the settler literature tradition. He likens the Indigenous women to the forest in which they live—"the women are disappearing into the forest. And then they are gone. Lost in the dark trees" (McKinnon 2017, 68)—implying that he and the other settlers are disconnected from their unfamiliar environs. Hawker is also quick to blame the surrounding habitat for his violent tendencies and inability to assimilate to the colonial frontier life:

What is this wild place that fills me with terrible purpose? It is some dark magic turning me away from my aim.... It is the *place* I must leave and those who inhabit it, those who work its dark magic. I must shoot the natives in the field.... Leave this forest and never return. (McKinnon 2017, 307; italics in original)

However, this colonial perspective of habitat is challenged by the story told from the perspective of Lola, a young Indigenous woman, and her experiences with settler descendants in 1900. Chronologically following the story of Hawker, Lola's narrative flips the novel from the perspective of a settler to that of an Indigenous person whose Country is the focus of the text. McKinnon is perhaps utilising these disparate stories to draw attention to Australia's violent history, and the perspectives of both the Indigenous people who endured it and the settler-colonisers who committed atrocious crimes. In this way, *Storyland* is not entirely complicit in constructing a traditional Australian settler narrative, as it attempts a more nuanced approach to the settler-colonial relationship with place. Langton argues that there is "this characteristic metaphor of the heroic lone white male or female throughout colonial history in texts [who is] relieved of any blame for their part in the demise of great civilisations" (1996, 18). *Storyland* resituates the hero or heroine in Australian settler narratives, portraying the domineering frontiersman character of Hawker as villainous and Indigenous teenager Lola as brave and heroic. She carries and shoots a rifle to protect those falsely accused of Jewell's murder (McKinnon 2017, 292) and plays a large role in finding the body of her missing friend. Her Aunty also displays unexpected influence when Jewell is found murdered: "There is such authority in her voice that people begin to step away" (McKinnon 2017, 298).

Lola is also particularly interested in the cultural history of the land and is influenced by her Aunty's thoughts on how their Indigenous heritage is entwined with the Lake Illawarra area. When speaking to local police office Constable Black about what horses can sense, Lola explains that

> … there are aspects of this world that we don't know about because we see in a particular way.… A sensing of a world that is different to our sensing … there are things science don't know. (McKinnon 2017, 288)

This suggests an understanding that the nonhuman can act in ways that are often unknowable to the human, and is a direct challenge to the Western notion that science is inherently able to reveal the perspectives of nonhumans. Steve Meredith, a Ngiyampaa Elder who was interviewed by Rose, speaks to this idea in a similar vein:

> What happens is because the fellow went to school, for birds, they're a bird expert, and then you get somebody went to school, they're a plant expert.… So that's why we say they pigeonhole things … [they think] they're higher than all that, they're above nature, and they tend to look down and study nature, like it's ants on the ground. But when you fall

asleep, eh, them ants they'll crawl all over you. They'll bite you, or sometimes they don't, but them ants might be carrying out their research then, on you. (2013, 99)

Meredith flips this idea of the human studying the nonhuman, which is also reflected in Lola's description of the horse as a being that is capable of its own sensing—a sensing that may be completely separate to that experienced by humans.

As much as this might indicate a deviation from traditional settler literature, it is worth noting that McKinnon's depiction of Indigenous characters as inherently in touch with nature and the nonhuman is possibly problematic. While McKinnon's novel may exemplify a culturally sensitive depiction of Indigenous connection to Country, she is also upholding the Western stereotype of First Nations people existing as one with the natural world. For example, Gordon Waitt critiques the Australian Tourist Commission's use of this stereotyping of Indigenous Australian people in the 1990s to represent "a primeval world, adventure in an unexplored frontier and ecotourism" to potential tourists (1999, 151). McKinnon does not engage in such stereotyping as explicitly, but her depiction of Lola as a kind of animal communicator reveals the still present trend within non-Indigenous Australian writing of engaging with the stereotyped idea of Indigenous "one-ness" with the natural world. However, the fact that Lola is the narrator of her own story lends some credence to the idea that McKinnon is attempting to portray connection to Country, rather than including a tokenistic Indigenous character that reflects the Western paradigm of considering Indigenous people as inherently in tune with the natural world.

4.5 The Rainforest and the Rural in Mireille Juchau's *The World Without Us*

The World Without Us is Mireille Juchau's third novel and was shortlisted for the Stella Prize 2016, longlisted for the Miles Franklin Literary Award 2016, and won the Victorian Premier's Literary Award 2016 for Fiction. Karen Hardy's description of the novel highlights its place within an emerging Australian settler literature tradition of responding to the colonisers' role in anthropogenic climate change:

Bees are dying, properties are being sold off to mining companies, there's a sense of despair amongst the farmers; it's a near future that rings too true. (2015)

The novel portrays the complex relationships within a small-town community—both the relationships amongst humans, and those between the human and nonhuman. Julieanne Lamond notes Juchau's exploration of the connection between people and the nonhuman, stating that

One of [the novel's] achievements is a really vivid evocation of a small town as a peopled landscape. Reading it, there is a strong sense of the interaction between the physical place … and the lived experiences of its inhabitants. (2015)

Evangeline and Stefan Müller have recently lost their young daughter, Pip, to leukaemia, the trauma of which has taken the greatest toll on Evangeline, or Evie, who seeks solace in an affair with out-of-towner, Jim. Meanwhile, the couple's eldest daughter, Tess, no longer speaks and Stefan discovers human remains on the family's farm, leading to a search to identify them and a confronting reliving of Evie's past. It is also later discovered that Stefan is not Tess' biological father. Meg, Tess' younger sister, is trying to make sense of the community's resistance to change in the face of environmental crisis. As described by Hardy above, the horror of environmental destruction, including the decimation of local European honeybee populations, hangs above the town's citizens as they deal with a myriad of personal issues. From survivalist Thomas Adam Tucker's clinical depression to Jim's inability to reconnect with his estranged wife, Sylvie, who lives in Sydney, the lingering horror of climate change and ecological devastation is a backdrop to and an instigator of people's troubles in this novel.

Juchau's portrayal of a native Australian rainforest habitat and the devastating impacts of climate change and environmental exploitation experienced in this place are prominent. In *The World Without Us*, the reader experiences the tragic consequences of human extractivism on the rainforest and the town nestled within it. The lives of those in the local community, including those who rely on farming practices that are impacting the natural environment, are drastically affected as the habitat falls to pieces around them. This emphasis on the dire consequences of extractivist practices and climate change and the narrative of a farming family learning to survive in an environment irrevocably altered by their own settler-colonial lifestyles sets this novel apart from other examples of Australian settler literature that instead glorify the work of those attempting to tame Australian habitats.

Nevertheless, when compared to McKinnon's text, the absence of any Indigenous Australian characters in *The World Without Us* is notable. Juchau advances this common and problematic aspect of Australian settler literature in her emphasis on the settler-coloniser's struggle to live on the land and the supplanting of any meaningful portrayal of Indigenous Australians. This resonates with Leane's comments regarding the prominence of the Australian bushman figure in twentieth-century Australian settler literature and the subsequent displacement of any appropriately portrayed Indigenous characters (2010, 32). As in the above section discussing *Storyland*, this concurrent engagement with and opposition to certain tropes of settler-colonial literature within the text will be explored in detail here.

Australian environmental activism has long been concerned with Australia's rainforest habitats and their exploitation by logging and mining companies. In recent years, the Great Forest National Park—a proposal that "Victorians create and add a new 355,000 hectares of protected forests to the existing 170,000 hectares of parks and protected areas in the Central Highlands of Victoria" (My/Forests Inc. 2017)—has been a major focus of various environmental political campaigns. The campaign to save the Leadbeater's possum, a critically endangered Australian marsupial, has also revolved around the protection of large mountain ash forests in Victoria's north-east in which the possum resides (Smith and Lindenmayer 1988). Megan C. Evans argues that little is known about how effective policy responses to deforestation have been in the past 40 years (2016, 145), which is concerning given the large-scale logging practices that have occurred in the native forests of south-eastern Australia over the past century (see Slezak 2018). Juchau does not endorse a colonial domination of these forests in her novel, but rather portrays the debilitating impacts of this domination and exploitation on human and nonhuman communities.

The habitat story embedded in *The World Without Us* therefore rests on the old-growth eucalypt forests that surround the Müllers' farm and town. These forests are, effectively, the primary nonhuman representation in this novel, ever-present and encircling the human and nonhuman experiences that occur throughout the narrative. The line between human and nonhuman is often blurred in the context of these forests; when the reader is first introduced to Evie through the eyes of Jim, she is "a pale column" seen "through the tawny trees ... by the Repentance River" (Juchau 2015, 3)—almost a tree herself. She is not an individual but rather a part of a habitat.

Interestingly, some of the most insightful commentary on the relationship between settler-colonisers and the local habitat is provided to the reader in the chapters focalised through Jim, an outsider to this community:

Somewhere on the mountain, axed wood starts groaning and splintering; his floor shudders when the trunks fall. He'll sometimes imagine ... that his cabin boards are sighing, he'll imagine the very sap of the wood stiffening in empathetic horror. (Juchau 2015, 92)

Juchau's depiction of the non-Indigenous relationship with an Australian native habitat both utilises and challenges a representation of landscape that is tied in with "symbols of conquest and colonial history" (Langton 1996, 12). *The World Without Us* delineates itself from the traditional settler-colonial narrative through Jim's discussion of environmental destruction as a reflection on un-belonging, rather than belonging. The horror lies not in an unknown wilderness, but rather in the destruction of this wilderness for Western-capitalist gain, as Jim describes. In this same chapter, Evie reads from Jim's notes, taken whilst on a field trip to the mountains with his students:

What is the message that wild animals bring ...? What is this message that is wordless, that is nothing more or less than the animal themselves—that the world is wild, that life is unpredictable in its goodness and its danger, that the world is larger than your imagination? (Juchau 2015, 92)

Jim's thoughts are ecocentric in nature, recognising that what is "wild" can be both dangerous and benevolent, and that the way humans perceive the world around them—in this case, the mountains—is limited. There are shifts throughout the novel between this kind of self-reflective prose—written by both Jim and Evie's daughter, Tess—and the dialogue between characters which is presented without quotation marks. This stylistic choice depicts human conversation as converging with the descriptive language that depicts the novel's setting. Rather than delineating what it is to be human from the experiences of the nonhuman, human conversation merges with the surrounding habitat. The personal writings of characters like Jim and Tess hint at something similar. Jim, or Mr. Parker, is also Tess' teacher and comments the following on her written musings: "A powerful account of place and memory. And where are you in the story, Tess? Which of the 'you's is you?" (Juchau 2015, 87). Tess' writing depicts the disappearance of her own self amongst the nonhumans and weather patterns of her home through her use of the second-person pronoun, "you", to describe her point of view. The horrific locust plagues and "the big rains" (Juchau 2015, 85) envelope Tess within her own story, further suggesting that the human characters of Juchau's narrative must be considered convergent beings existing with the surrounding natural world.

Jim's wife Sylvie, the one character of the novel who resides outside of this community in the nearest city (Sydney), is mistaken about the reality of living in a community that is torn between "farm or mountain" (Juchau 2015, 81). She tells her soon-to-be estranged husband Jim that moving away from the city constitutes an "escape"—"to clean air and mountains, rivers that you can probably even drink from" (Juchau 2015, 81). Such a viewpoint is indicative of the way that other authors within the settler literature tradition have portrayed the so-called wilderness areas of the place-to-be-conquered. Stemming from the pastoral tradition[4] and the latter Romantic tradition,[5] these perspectives of the wild as inherently pure and beautiful ignore that, even if beautiful, these are also places of colonial violence and ecological destruction. Sylvie sees the relationship between her husband and the rainforest as a one-sided, voyeuristic one—Jim is living *on* the land, and not *amongst* it. As the intertwining of the human and nonhuman show, this is far from the case amongst many of the community's residents, particularly for Jim as indicated by his musings above.

Tess, who has not spoken since the death of her sister, often reflects on her place within a larger, nonhuman world and the influence of the climate on her family's way of life. Mr. Parker encourages Tess to ponder the following questions:

How is [the weather] connected to your family? How is the atmosphere part of who you are? ... When [the climate is] unpleasant, when it's raw, dirty and clammy, how do we get through? (Juchau 2015, 83)

These ideas are depicted as being at the heart of much of what Tess considers throughout the novel, and are reflected in her writings that highlight her struggle to comes to terms with her own identity (as described above, she consistently refers to herself as "you" rather than "I" in her written work). This reflects a nuanced take on settler-colonial belonging in native habitats; an unsettled belonging that questions one's place in it. However, one way in which Juchau is perhaps complicit in a

[4] For a discussion of the history of pastoralism and its representation of the nonhuman, see Scheese (2002).

[5] For a summary of this, see Cronon (1996).

tradition of Australian settler literature is the links she draws between the personal problems of characters and environmental issues. Whilst this may highlight the inextricable connections that exist between human and non-human issues, it also plays into the idea that the nonhuman experience is allegorical for a human one. Following the above questions from her teacher, Tess realises that "she knew he wasn't talking about the weather.... How could her teacher know her mind, and what her thoughts were running from?" (Juchau 2015, 83). This link elevates Tess' personal struggles with grief and family above the rapidly growing issue of climate change.

Despite this, much of Juchau's text engages with the coloniser's relationship with the native habitat in a more novel way than most traditional settler literature. Early in the narrative, Jim remarks that "the fey place names—Rainbow Hill, Naiad Gorge, Moonbeam Falls—had replaced all the indigenous names, and were absurd" (Juchau 2015, 4). Any feeling of belonging he, or indeed any settler-coloniser, might experience in this habitat is therefore inauthentic: "He could pretend, in the surround sound of wind, water, leaf and bird, that the forest was pristine, that he was born to it and belonged" (Juchau 2015, 4). For Jim, this sense of belonging is all pretend, especially in a place where the local habitat is being decimated by mining (one of many destructive consequences of colonisation). Tess' younger sister Meg views this as a kind of ferocious decimation and she recognises the irony of the township preventing supposedly aggressive, pro-environmental protests whilst allowing loggers to be "violent to the trees" (Juchau 2015, 72).

The attempts of people like Evie and local protestors to prevent further devastation are portrayed as overly optimistic and inevitably futile in the face of settler-colonial extractivism. Meg notes that, despite her mother's best efforts to assist young protesters, fracking had gone ahead in the forest, resulting in "twenty active wells on the mountain, topped by metal rigs, and five ponds of chemical waste" (Juchau 2015, 73). Although the protests are futile, the novel's critique of settler-colonial extractivist practices represents a challenge to a vital trope of many works of Australian settler-colonial literature: the conquering and exploitation of the land as an empowering means by which settler-colonisers seek belonging. Potter contends that within settler-colonial culture, "virtue and legitimacy are conferred by the entrepreneurial utilization of fallow resources" (2019, 19). While speaking here of the exploitation of the land for Anglo-agricultural purposes, Potter's point can be extended to Juchau's condemnation of mining, fracking and logging (as well as farming) throughout *The World Without Us.*

4.6 A Critique of Fetishised Settler-Colonial Rurality

Juchau's work also highlights the problem of fetishising rural living in settler-colonial landscapes and the resultant nostalgia for a life that can no longer exist. This is a problematic way to live in a time of climate crisis, as there is a pervasive assumption that things will remain as they always have been or that for life to be worth living, things must return to what they once were. *The World Without Us* portrays a community that is simultaneously trying to come to terms with its own troubled history and move into a future of almost unimaginable difference. Climate change presents an opportunity to both embrace country living in the sense that it brings people closer to nature and challenge the elements of it that subscribe to a settler-coloniser way of life, such as the Euro-pastoral landscapes that are responsible for so much habitat clearing across Australia (Slezak 2018).

Classic settler-Australian novels such as Jeannie Gunn's *We of the Never Never* (1908), Colleen McCullough's *The Thorn Birds* (1977) and Tim Winton's *Cloudstreet* (1991) celebrate these rural landscapes as places that, despite hardships, solidify settler belonging (particularly when compared to the city); in doing so, they forego an acknowledgement of the destructive settler-colonial practices that created and are now irrevocably changing these celebrated settler places. Juchau critiques this celebration of Australian country living throughout her novel. Thinking about how "a city bee" is arguably healthier than a country one, Tess ponders how "country life is what damages you: the unreported mine leaks, the rolled tractors, the pesticides" (Juchau 2015, 228). This suggests that Australia's stereotypically pure rainforests and their human communities are more at risk of environmental degradation than cities. Are these rainforests therefore "shadow places" (2008)? They provide both "material and ecological support" (Plumwood 2008, 139) to humans, and are often denied—either intentionally or unintentionally—as places of cultural and ecological significance by those in positions of power to protect them.[6] Juchau's novel therefore redirects readers' attention to rural Australian

[6] In March 2019, *Guardian Australia* reported on one instance of this, explaining how a "damning logging report" had revealed that the Victorian governmental body responsible for timber harvesting regulation had been largely ineffective, allowing VicForests, a government-backed logging company, to self-regulate its logging of native forests (Cox 2019).

communities as places of environmental destruction and damaging settler-colonial traditions, rather than places that facilitate wholesome country living.

They are also often places of divisive environmental politics, residents torn between protecting the environmental beauty that they hold dear and ensuring that they maintain their livelihoods when faced with the pressure of living more ecologically minded lifestyles. Meg notes that her father "loves the land and, still, he's had the Clear Energy men out to the farm because he wants to get both sides of the story" (Juchau 2015, 74). As the fracking company responsible for the destruction of local forest, Clear Energy represents settler-coloniser extractivism in one of its most extreme forms (see Batley and Kookana 2012), while Stefan's farm is presented as less invasive—it is something that he loves and associates with the native habitats that encircle it. It is, of course, problematic for Stefan to ignore the impact that his own farming has had on local habitats and ecosystems, and Juchau suggests that Stefan, as someone who loves both his farmland and the surrounding forest, still remains open to the extractivist industries that will inevitably impact not just native habitats but his own farm and livelihood as well. Stefan's values may seem incongruous with one another, but they speak to a broader problem within settler-colonial society of supporting certain industries on the premise of employment and profit in spite of the predicted future impacts on the livelihoods and well-being of settler-colonisers themselves.

Significantly, the novel culminates in a direct confrontation between city and country when Tess and her fellow students from the markedly named "River School" meet the visiting "city girls" from "Sancta Sophia" (Juchau 2015, 267). They "had come for the reforesting—not just for their future, but for the generations" (Juchau 2015, 268). The city students are portrayed in stark contrast to Tess, and are impressed by her family's farming background and unusual history (Juchau 2015, 272–273). This meeting suggests that the local community mindset is expanding with the recognition that the consequences of local environmental devastation spread further than the town and surrounding habitat itself. However, the awkward nature of this meeting and the emphasis on the differences between city and country suggest that there is much more work to be done, on both sides. In some ways, Tess is a new kind of heroine for Australian settler literature—not one who "who face[s] the adversary" of Australia's wilderness as Langton describes (1996, 11), but one who faces a very different adversary: the denialism of a rural town that refuses to enact real action on local environmental crises that affect both the local human community and the nonhuman habitat that defines this community. In this sense, Juchau's novel does not fit within the more traditional scope of Australian settler literature, as it challenges the assumption that Australia's native habitats exist to be exploited and are incongruous with settler-colonial life. Rather, *The World Without Us* explores how these habitats might foster opportunities to better understand settler-colonial identity as something that is constantly changing alongside the identities of these habitats—not as a stable and separate way of being.

4.7 Memory and Habitat

Whilst Juchau's novel does not explore the relationship between time and habitat in the same way that McKinnon's work does, *The World Without Us* includes a significant focus on memory and its influence on human interactions with the nonhuman world. One of the novel's primary narrative arcs follows Evie as she comes to terms with her past life in the town's now-defunct hippy commune, "The Hive" (Juchau 2015, 172). Significantly, much of Evie's past is viewed through the lens of other characters—her family, and other members of the local community. Rarely does the reader experience a focalisation of past and present events through Evie herself. In a chapter midway through the novel, the reader is finally exposed to Evie's memories of the night many years ago when The Hive caught fire and she attempted to escape, only to be kidnapped by two local boys in their van (Juchau 2015, 172). Both the fire in Evie's past and the flood in her present symbolise a resetting; an opportunity for her to break free from her "hippie homeschooled" (Juchau 2015, 173) past and her grieving present following the death of her daughter. She believes "… the fire was blessed. Maybe even a cleansing…. Wasn't she newly forged in that fire, hadn't she become another person entirely?" (Juchau 2015, 173).

It is during the flood in Evie's present—amongst "long pillars of rain, a ledge of slate-coloured sky" (Juchau 2015, 245)—that she gives birth to her son, trapped and alone in the family's home. This birth is another resetting, the beginning of a more concrete stage in her and her family's life as "the small boy governs their home" (Juchau 2015, 264) and Evie begins to acknowledge her past, including the fact that Tess does not know her biological father. These events—the flood and the fire—therefore intertwine with human memory in significant ways, providing a clean slate for this settler-colonial Australian family and wider community to come to terms with the parts they have played in local environmental destruction and how a renewed sense of belonging may assist in acting on climate change. As Potter posits in her work on non-Indigenous belonging in postcolonial Australia, it may be an opportunity to search for a new belonging that no longer relies on the colonial domination of nature (2013, 33).

While endorsing the search for a new form of non-Indigenous belonging, Juchau's narrative does not seek to redeem the settler-colonial project. The conclusion to her work instead suggests a shift away from the destructive extractivist industries introduced to Australia by colonisers:

"So much has gone," my mother says again. And I know what she's thinking about, as her eyes search the scarred ridges and crests, the mists or dusts furling around the remaining forest trees—and I know who. "Yes. Pip is gone," I say… "But we are here. Look, here we are." (Juchau 2015, 293)

Pip's death marked a significant change in the way the Müllers lived their lives, and so too does this final moment narrated by Tess, who is finally speaking again. There is a recognition here that the family's lives must change even more, incorporating the lessons learned from the decimation of local habitat. There is a recognition that the settler-coloniser is still here, living on stolen, damaged land, but rather than accepting this, the Müller family, particularly Tess, are beginning the work of challenging the settler-colonial project in order to combat ecological crisis in their town. Notably, the aptly named "Repentance River" (Juchau 2015, 3), which is described in the novel's opening chapter and foreshadows this conclusion, alludes to the struggle of Australia's settler-descendant society in coming to terms with the violence of the past and the unstoppable threat of climate change that they have played a part in bringing about.

Similarly, some passages also allude to the symbology of European trees in the town—a common sight in many rural Australian towns, even those located amongst native rainforest. Meg observes forest leaves which appear "scarlet and gold on the black tarmac", as well as "winter branches blown bone-clean" (Juchau 2015, 70). The likelihood is that the leaves and clean "winter branches" are those of European trees—not eucalypts and other native trees which make up the majority of Australian rainforest and do not shed their leaves in the autumn and winter. This intrusion of non-native foliage emphasises the lack of belonging that Meg and her family often experience throughout the text; they are strangers to the place in which they have spent most of their lives. Using the themes of climate change and environmental exploitation, Juchau posits that an opportunity exists for "a re-articulation, and indeed intensification" of the discourse surrounding non-Indigenous belonging in postcolonial Australia, as Potter describes (2013, 30). Climate change may be seen as an "unsettlement of non-Indigenous dominion", offering a chance for non-Indigenous Australians to reconsider the impacts that their dominance of native habitats and species has had and to reposition themselves to embrace a sense of belonging that does not rely on "identity discourse that fixates on conquest and heroic progress" (Potter 2013, 33).

In contrast to this, though, *The World Without Us* remains largely silent on Indigenous place and belonging. One could argue that Juchau's text upholds the notion that settler-colonisers may struggle with their sense of belonging, as the Müllers do, but that, as Potter posits, they still inevitably learn from the trials they have suffered in the pursuit to "redeem or restore" alien landscapes (2019, 18). Stemming from a "pastoral retreat" tradition (Potter 2019, 18), this pursuit rings true in the context of Juchau's narrative, despite the fact that the Müller family and others within their town eventually come to terms with the impact of their unsustainable settler-colonial practices.

4.8 Conclusion

Taking into account the past, and still largely present, tropes of much Australian settler literature as glorifying an Australian wilderness that settler-colonisers both exploit and contain, the novels analysed in this chapter provide a rethinking of these traditions in their depictions of habitat stories. Discussion of Australian habitats by non-Indigenous scholars has long been confined to science and environmental management, and so this analysis of McKinnon's and Juchau's work paves the way for more complex literary interpretations of the ways in which Australian settler authors portray habitat in their works. McKinnon's experimentation with a nonlinear narrative and her nuanced exploration of the Lake Illawarra habitat and its inhabitants invites readers to consider the human and non-human stories that occur here as a means of understanding the modern environmental crisis, and how settler-colonisers must take responsibility for the devastation of the Australian natural world and the horrific violence and dispossession committed against Indigenous Australian peoples. *The World Without Us* also presents habitat as playing a vital role in the story of the Müller family, and yet its destruction still occurs, providing a lesson for the novel's non-Indigenous characters as they grapple with the consequences of extractivist technologies on local habitats, but also their own way of life as settler-colonisers.

However, it is important to recognise that *Storyland* and *The World Without Us* still perpetuate some of the more traditional tropes of Australian settler literature. The absence of Indigenous characters in Juchau's work and the primary focus on non-Indigenous experience in McKinnon's are two examples of this, as well as the shift between habitat as an agential being with its own stories to a representational entity that is used by the authors to symbolise the settler-colonial struggle for belonging on stolen land. Nevertheless, their nuanced take on the role that habitat plays in postcolonial Australia's approach to ecological crisis makes these texts significant markers of a developing trend in settler literature that is seeing authors grapple with the colonial impact on the nonhuman in generative ways.

References

Armstrong, Philip. 2008. *What Animals Mean in the Fiction of Modernity*. London: Routledge.

Batley, Graeme E., and Rai S. Kookana. 2012. Environmental Issues Associated with Coal Seam Gas Recovery: Managing the Fracking Boom. *Environmental Chemistry* 9 (5): 425–428.

Brewster, Anne. 2015. *Giving This Country a Memory: Contemporary Aboriginal Voices of Australia*. Amherst, New York: Cambria Press.

Burarrwana, Laklak, Ritjilili Ganambarr, Merrkiyawuy Ganambarr-Stubbs, Banbapuy Ganambarr, Djawundil Maymuru, Sarah Wright, Sandie Suchet-Pearson, and Kate Lloyd. 2019. *Song Spirals: Sharing Women's Wisdom of Country through Songlines*. Crows Nest: Allen & Unwin.

Carter, Paul. 2009. *Dark Writing: Geography, Performance, Design*. Honolulu: University of Hawai'i Press.

Cox, Lisa. 2019. Damning Logging Report Finds Victorian Department Neither "Effective nor Respected". *Guardian Australia*, March 25. https://www.theguardian.com/australia-news/2019/mar/25/damning-logging-report-finds-victorian-department-neither-effective-nor-respected. Accessed 29 Jan 2025.

Cronon, William. 1996. The Trouble with Wilderness: Or, Getting Back to the Wrong Nature. *Environmental History* 1 (1): 7–28.

Davis, Heather, and Zoe Todd. 2017. On the Importance of a Date, or Decolonizing the Anthropocene. *An International Journal for Critical Geographies* 16 (4): 761–780.
De Cristofaro, Diletta. 2018. "Time, No Arrow, No Boomerang, But a Concertina": *Cloud Atlas* and the Anti-Apocalyptic Critical Temporalities of the Contemporary Post-Apocalyptic Novel. *Critique: Studies in Contemporary Fiction* 59 (2): 243–257.
De Cristofaro, Diletta. 2020. *The Contemporary Post-Apocalyptic Novel: Critical Temporalities and the End Times*. London: Bloomsbury Academic.
Deng, Junjie, Brian G. Jones, Kerrylee Rogers, and Colin D. Woodroffe. 2018. Wind Influence on the Orientation of Estuarine Landforms: An Example from Lake Illawarra in Southeastern Australia. *Earth Surface Processes and Landforms* 43: 2915–2925.
Donovan, Josephine. 2016. *The Aesthetics of Care: On the Literary Treatment of Animals*. New York: Bloomsbury.
Evans, Megan C. 2016. Deforestation in Australia: Drivers, Trends and Policy Responses. *Pacific Conservation Biology* 22: 130–150.
Grenville, Kate. 2005. *The Secret River*. Edinburgh: Canongate.
Griffiths, Billy. 2018. *Deep Time Dreaming: Uncovering Ancient Australia*. Carlton: Black Inc.
Gunn, Jeannie. 1908. *We of the Never Never*. London: Hutchinson.
Hardy, Karen. 2015. The World Without Us: Mireille Juchau Explains Why Climate Change Matters. *The Sydney Morning Herald*, September 3. https://www.smh.com.au/entertainment/books/mireille-juchau-and-the-world-without-us-20150902-gjdheu.html. Accessed 29 Jan 2025.
Juchau, Mireille. 2015. *The World Without Us*. London: Bloomsbury Publishing.
Kwaymullina, Ambelin. 2020. *Living on Stolen Land*. Broome: Magabala Books.
Lamond, Julieanne. 2015. The Atmosphere We Live In: The World Without Us. *Sydney Review of Books*, October 9.
Langton, Marcia. 1996. What Do We Mean by Wilderness?: Wilderness and Terra Nullius in Australian Art [Address to The Sydney Institute on 12 October 1995]. *The Sydney Papers* 8 (1): 10–31.
Leane, Jeanine. 2010. Aboriginal Representation: Conflict or Dialogue in the Academy. *The Australian Journal of Indigenous Education* 39 (Supplement): 32–39.
Leane, Jeanine. 2013. Tracking Our Country in Settler Literature. *Journal of the Association for the Study of Australian Literature* 14 (3): 1–17.
LeMenager, Stephanie. 2017. Climate Change and the Struggle for Genre. In *Anthropocene Reading: Literary History in Geologic Times*, ed. Tobias Menely and Jesse Oak Taylor, 220–238. University Park: Pennsylvania State University Press.
Mandel, Emily St John. 2014. *Station Eleven*. New York: Alfred A. Knopf.

Marshall, James Vance. 1959. *The Children*. London: Michael Joseph.
McCann, Andrew. 2006. The Literature of Extinction. *Meanjin* 65 (1): 48–54.
McCullough, Colleen. 1977. *The Thorn Birds*. New York: Harper & Row.
McKinnon, Catherine. 2017. *Storyland*. Sydney: HarperCollins Publishers.
Mitchell, David. 2004. *Cloud Atlas*. London: Sceptre.
Morris, Linda. 2017. Catherine McKinnon's *Storyland*: A Work Transcending Time. *The Sydney Morning Herald*, March 9. https://www.smh.com.au/entertainment/books/the-stories-of-us-by-catherine-mckinnon-review-20170224-guk5t8.html. Accessed 29 Jan 2025.
My/Forests Inc. 2017. *Place*. Great Forest National Park. https://www.greatforestnationalpark.com.au/park-plan.html. Accessed 29 Jan 2025.
Oerlemans, Onno. 2004. *Romanticism and the Materiality of Nature*. Toronto: University of Toronto Press.
Pascoe, Bruce. 2014. *Dark Emu: Aboriginal Australia and the Birth of Agriculture*. Broome: Magabala Books.
Plumwood, Val. 2008. Shadow Places and the Politics of Dwelling. *Australian Humanities Review* 44: 139–150.
Potter, Emily. 2013. Climate Change and Non-Indigenous Belonging in Postcolonial Australia. *Continuum* 27 (1): 30–40.
Potter, Emily. 2015. Programme: Dreaming, Timekeeping and Becoming. In *Assembling the Centre: Architecture for Indigenous Cultures*, ed. Janet McGaw and Anoma Pieris, 100–120. London and New York: Routledge.
Robin, Libby. 2007. *How a Continent Created a Nation*. Sydney: University of New South Wales Press.
Rose, Deborah Bird. 2013. Val Plumwood's Philosophical Animism: Attentive Interactions in the Sentient World. *Environmental Humanities* 3 (1): 93–109.
Rowlands, Mark. 2009. *The Philosopher and the Wolf*. London: Granta.
Scheese, Don. 2002. *Nature Writing: The Pastoral Impulse in America*. New York and London: Routledge.
Slezak, Michael. 2018. "Global Deforestation Hotspot": 3m Hectares of Australian Forest to Be Lost in 15 Years. *Guardian Australia*, March 5. https://www.theguardian.com/environment/2018/mar/05/global-deforestation-hotspot-3m-hectares-of-australian-forest-to-be-lost-in-15-years. Accessed 29 Jan 2025.
Smith, Andrew P., and David Lindenmayer. 1988. Tree Hollow Requirements of Leadbeater's Possum and Other Possums and Gliders in Timber Production Ash Forests of the Victorian Central Highlands. *Australian Wildlife Research* 15 (4): 347–362.
Stadler, Jane, Peta Mitchell, and Stephen Carleton. 2015. *Imagined Landscapes: Geovisualizing Australian Spatial Narratives*. Bloomington & Indianapolis: Indiana University Press.
Tranter, Aunty Joan. 2013. Indigenous Country Explained. *Education* 94 (7): 11.

Waitt, Gordon. 1999. Naturalizing the "Primitive": A Critique of Marketing Australia's Indigenous Peoples as "Hunter-Gatherers". *Tourism Geographies* 1 (2): 142–163.

Whyte, Kyle P. 2018. Indigenous Science (Fiction) for the Anthropocene: Ancestral Dystopias and Fantasies of Climate Change Crises. *Environment and Planning E: Nature and Space* 1 (1–2): 224–242.

Williams, Nicholas S. G., Adrian Marshall, and John W. Morgan, eds. 2015. *Land of Sweeping Plains: Managing and Restoring the Native Temperate Grasslands of South-Eastern Australia*. Clayton: CSIRO Publishing.

Winterson, Jeanette. 2007. *The Stone Gods*. Orlando: Harcourt.

Winton, Tim. 1991. *Cloudstreet*. Carlton: McPhee Gribble.

CHAPTER 5

Unstable Ground and the Abiotic Nonhuman in the Regional Australian Novel

5.1 Post-industrialism and Australian Regionalism

The Australian regional novel is what we might consider a key genre of Australian literature—an influential form that has been used by white settler authors to reinforce colonial mythos about the settler connection and right to land. Philip Steer describes the progress of the (British) seasons as sitting at the heart of British realism's regional novel (2021, 371). The regional novel of colonial Australia was no different, although this of course involved "the transplanting of British weather expectations" onto the vastly different seasons and environments of the Australian continent in an effort "to replicate Victorian society through the establishment of a pastoral economy" (Steer 2021, 372). Attending more closely to climate and the nonhuman in the novel means looking to how geographies and, by extension, regions are "an active force, that pervades the literary field and shapes it in depth" (Moretti 1998, 3, cited in Steer 2021, 372).

The novels discussed in this chapter seriously question the validity of the British regional mythos in Australia by subverting and at times satirising tropes of the regional novel. *Dyschronia* and *Australiana* both highlight the impacts and the inevitability of continuing extreme weather events by satirising the attitude conveyed in traditional regional novels that these things are blue-moon events, a drop in the ocean of supposed climatic predictability—a notion carried over from colonialism that

R. Fetherston, *Theorising the Postcolonial Eco-Novel*, Literatures, Cultures, and the Environment,
https://doi.org/10.1007/978-3-032-04466-2_5

Australian environs are tameable and if a community can just get through the next drought or flood, then things will return to "normal". In *Dyschronia*, the townsfolk experience "a hot morning, strange for the season" (Mills 2018b, 11), while in *Australiana*, the townsfolk experience what they perceive to be the "worst drought in history" (Kassab 2022, 64). As Steer argues, the realist form, seen in the light of Anthropocene and climate change discourse, "brings to the fore long-standing questions of realism's commitment to probability" (2021, 375). While Steer argues that the British regional novel actually situated extreme weather events within their own pattern of predictability—often based on local climate knowledge and folklore that revealed some kind of regularity of extreme weather events (2021, 375)—the Australian regional novel of the nineteenth century was dealing with an altogether different beast. Drought in particular brought an "ontological challenge" (Steer 2021, 377) to the colonist. "Neither the cause nor the extent of a drought appeared to be predictable", Steer describes, "nor was it apparent if drought should be seen as an anomalous event or an inherent part of the climate" (2021, 377).

This chapter offers a discussion of two eco-novels which I also define as part of the Australian regional literary tradition—a tradition that interrogates the inner lives and conflicts of Australians living in regional and rural areas. The first of these is Jennifer Mills' *Dyschronia* (2018): a speculative literary novel which emphasises the post-industrial degradation of a regional town in South Australia through its depiction of what I term the "abiotic nonhuman" (in this case, a monstrous plaster sculpture of a squid and various products of the asphalt industry) and its subversion of settler understandings of climate catastrophe. Similarly, Yumna Kassab's *Australiana* (2022) portrays another regional "everytown" of inland Australia (possibly in New South Wales) upset by ecocatastrophe, this time in the form of extreme drought and flooding, and emphasises the role of nonhuman objects such as stones, wool and agricultural equipment in both validating and unsettling different communities' connection to place.

Regional literature in Australia was not overtly recognised in literary discourse until the late 1970s—a time when stated-based literary anthologies were being published across Australia, including several focused on Western Australian literature (Henningsgaard 2007, 13). Regionally focused literature has long been marginalised within the broader framework of Australian national literature, which itself has been marginalised in the even broader international literary scene (Henningsgaard 2010, 67), particularly in the northern hemisphere. Despite this marginalisation, Per

Henningsgaard acknowledges the role of Australian regional literature in resisting and critiquing globalised processes (2010, 69). It therefore seems prudent to further investigate the truth of this in the more specific context of ecological crisis.

The tradition of the regional novel is reliant on the traditions of production and industry stemming from nineteenth-century colonialism. Paradoxically, the future of the kinds of Australian towns that the regional novel describes is reliant on the post-industrial ideologies that have, of course, also emerged from colonial industrialism, but that now assert a very different, albeit related, kind of capitalist-colonial domination of people and the more-than-human. A consideration of post-industrialism in the context of the Australian regional novel is therefore key. As Daniel Bell described it in 1973, the "post-industrial" refers to a society in which "relation to the instruments of production no longer decides dominance or power or privilege in society" (57); in other words, a society "dominated by an educated professional-technical science class" (105). This society is still ruled by the tenets of industrialism, particularly the use of alienation to convert "people and things" into "mobile assets" (Tsing 2021, 5). Such assets are "removed from their life worlds in distance-defying transport to be exchanged with other assets from other life worlds, elsewhere" (Tsing 2021, 5).

Understanding the role of the nonliving nonhuman—many of which are these "things" produced through and entangled in capitalist systems of production—in dismantling such modes of production is also key. In their work on multispecies ethography, Thom van Dooren and Deborah Bird Rose explain that nonliving entities are just as important in understanding the lifeways and stories of the more-than-human. Significantly, an attentiveness to an abiotic nonhuman "undoes any sense of a firm border between the living and the nonliving and throws into question certainty about what this distinction might even mean" (2016, 87). It is also not enough to attend only to literary representations of the living nonhuman when the nonliving or the abiotic plays such a significant role in the Australian regional novel. Objects and infrastructures have always loomed large in the Australian regional novel, particularly those relating to agriculture and the alteration of lands and waters in the name of nation-building. The post-industrial age brings an added complexity to how human–nonhuman relations are configured in the Australian regional novel, and it is this complexity that I am chiefly concerned with in this chapter.

An emphasis on the nonliving nonhuman is also needed in any ecocritical forum if we are to properly attend to arguably the most significant human–nonhuman relationship in the context of the Anthropocene—that between humans and fossil fuels. As Kathryn Yusoff phrases it, the "matter-energy" of (nonliving) Carboniferous fossils is what has fuelled humanity's "ascendance to an inhuman planetary power" and made possible "the political formations of late capitalism" (2013, 780). This fact alone urgently necessitates an increased emphasis on the influence of the abiotic nonhuman in environmental and ecocritical discourse. Both novels discussed in this chapter challenge the linear progress narrative of industrial and post-industrial capitalism, with the abiotic nonhuman suggesting the possibility of human–nonhuman connection as much as it reflects the radical upending of industrial development and wealth in regional and rural Australia. Elizabeth Smyth is one literary scholar who has identified the significance of the abiotic nonhuman in Australian regional fiction, arguing that farm machinery in the Australian farm novel has the capacity to challenge dominant settler-colonial ideologies about land use (2022). Smyth's work, though, is focused on machines that are imbued with "magical agency"—something that is "rarely encountered in Australian literary fiction" (2022, 3). Comparatively, the characterisation of the nonliving nonhuman that I discuss positions these objects as agential through their real-world materiality, rather than being granted agency via magical realism.

Attending to the nonliving nonhuman also means attending to the trivial, everyday objects that reflect what in reality are the very serious, far-reaching impacts of the climate crisis, particularly how it is linked to (post) industrial capitalism. The emphasis that these texts place on community and the everyday human and nonhuman experiences in the context of serious environmental devastation reflects what Stephanie LeMenager says about the place of climate change within certain literary genres:

> Climate change represents … an assault on the everyday. This means that it will register in those genres intended to query and explore what is understood as probable, cyclical, and even trivial experience. (2017, 221)

Many Australian ecofiction novels explore just that, especially in regard to the everyday, cyclical nature of both nonhuman and human life. These texts often do not approach climate change as an external force or a distinct event, but rather embed the impacts of climate change, both the obvious and the subtle, within their narratives, reflecting on the past, present and future and the notion that human and nonhuman history repeats itself in ways that are both expected and unexpected.

The uncertainty of the colonial landscape has heightened anxieties around this "assault on the everyday" (LeMenager 2017, 221). Relevant to this is Emily Potter's consideration of "the colonial obsession with stabilized ground" (2019, 103). Potter defines the concept of ground as relating to both nonliving matter and energy:

> ...it is rock, but it is also atmosphere; it is relational and it is always in movement, but it guarantees no certain futures, nor finalized relations. It can hide but also reveal.... And it suggests, through this, an experience of living in history, in the midst of things unfolding. (2019, 103–104)

Dyschronia and *Australiana* attend to this unfolding through their engagement with the abiotic nonhuman, challenging the supposedly stable ground of colonial settlement in contemporary Australia.

5.2 Time Slips and the Industrialised Nonhuman in Jennifer Mills' *Dyschronia*

Shortlisted for the Miles Franklin Award in 2019, Mills' speculative novel *Dyschronia* made a significant impact on Australian critics. Gretchen Shirm describes it as a "mesmerising take on the future of Australian coastal life" (2018), while James Bradley, a fellow author of Australian climate fiction, believes that in *Dyschronia*, Mills finally departs from "the niceties of literary realism altogether, venturing instead into the unsettled—and unsettling—hinterland of the fantastic" (2018). *Dyschronia* is indeed a difficult text to categorise. Mills employs distinctly dystopian themes in her realisation of a near-future ecocatastrophe which turns Clapstone, a fictional coastal South Australian town, into a ghost town. Interestingly, though, Mills herself describes the novel as "not a true dystopia, although it flirts with the form" (2018a). She believes that it explores "the question of human accountability to and for the future—a question that remains for me unresolved" (2018a). She also says in regard to reader response to the text that

> ...*Dyschronia* has sometimes been confused by its themes, taking its articulation of the mechanics of imagining the future as itself a prediction of that

> future.... But the axis of optimism/pessimism misses the real issue: the role of human agency in ***determining*** the future, both individually and collectively. (2018a)

Like other novels discussed in this book, *Dyschronia* grapples with the complex idea of agency in the climate crisis, both in terms of the nonhuman having agency and humans remaining ignorant of their obvious agency in a world being slowly, yet surely consumed by the effects of climate change. Although aspects of the novel might be defined as speculative, Mills does not present the ecocatastrophe experienced in Clapstone as a future event; it is an event that could happen as soon as today or tomorrow, the novel grounded in the here and now and reflecting the impacts of post-industrial social and economic shifts on small regional towns in Australia today. *Dyschronia* is particularly concerned with how post-industrialism—represented by a failing asphalt industry—coupled with environmental disaster spells the end of settler-industrial relations with the land. What is left in its wake is the corporate-speak of shallow tourism that relies on healthy ecosystems and communities to thrive—ecosystems and communities impacted by the past 250 years of colonial production and, now, the absence of such production and its replacement with something even more unsustainable. The abiotic nonhuman plays a central role in this story, representing both the product of post-industrial exploitation of the land and water, and the potential for a more empathetic dialogue between the human and nonhuman.

The beginning of the novel sees Clapstone locals wake one morning to find that the sea has disappeared, leaving "distant corpses" (Mills 2018b, 13) of marine fauna decomposing on the exposed seabed. Interestingly, this aspect of the novel—the receding seawaters—is in many ways "a reversal of the usual inundation narratives" of cli-fi (Hughes-D'Aeth 2023, 294) that tend to focus on the impacts of rising sea levels. At its heart, *Dyschronia* is the story of a mother and daughter attempting to escape a problematic small-town country life in a period of immense environmental and social upheaval. The narrative centres on Sam, a young woman coming to terms with her ability to see into the future, and her relationship with her mother, Ivy, a woman of questionable morality who is torn between the well-being of her daughter and the opportunity to profit from Sam's ability or, more accurately, illness. Like Oblivia in *The Swan Book*, who suffers from a "diseased mind" (Wright 2014, 5), Sam suffers from an illness that defines much of her experience throughout the narrative. Out-of-towner Ed soon hears of Sam's illness—this ability to sometimes see the events of the future and in turn suffer debilitating migraines, which is termed "dyschronia" (Mills 2018b, 65) by doctors—and begins a plan to exploit Clapstone locals in a development scheme destined for disaster.

The imaginary governmental "DSC", or "Department of Sustainable Communities" (Mills 2018b, 148), ironically does little to assuage the fears of those left in Clapstone by the end of the novel, wanting less to develop a sustainable community out of the now-defunct town and more to forget it ever existed in the first place. Mills herself admits to drawing inspiration for the DSC from "'commonsense' newspeak like 'practical reconciliation'" (2018a), suggesting that the way settler-colonisers speak about sustainability and the natural environment needs an ecocentric realignment that reflects both the accuracy and urgency of the modern environmental crisis. The regional setting of the novel also draws associations between the stereotype of Australian small-town ignorance and this regional community's lack of response to environmental catastrophe. Significantly, the lack of meaningful connection between the town's settler-Australians and the nonhuman leads not only to an ecological crisis but also to a deep, irreversible disturbance of settler-colonial notions of home and livelihood.

The chronology of events in the novel also leaves much open to interpretation. Mills states that she was interested in exploring "the dislocations of action and consequence that occur in a dyschronic narrative timeline" (2018a). I argue that this exploration of time also pertains to the human understanding and relationship with the abiotic nonhuman, as it reflects humans' connection to the geologic nonhuman through evolution and deep time. While the living nonhuman is a large presence in the narrative—primarily in the form of the living and dying cuttlefish that protagonist Sam is connected to—the abiotic nonhuman is an equally significant figure here, represented by a larger-than-life squid sculpture that punctuates Sam's visions. I will discuss both here because, as will become evident, the living cuttlefish and the nonliving squid sculpture are intimately connected to one another via Sam, both playing a role in the unsettlement of the Clapstone township.

The reader of this novel experiences a deep, but often troubled, relationship with the nonhuman primarily through the eyes of Sam. Like many of the novels discussed in this book, *Dyschronia* portrays the developing connection between a human and the nonhuman—in this case, between Sam and the Australian giant cuttlefish, whose population has been devastated. The opening line of the novel foreshadows what the reader later learns about the relationship Sam forms with the cuttlefish: "At first Sam thought the scratching was an animal" (Mills 2018b, 1). The reader is told that it cannot be a possum as there are "no possums any more" (Mills 2018b, 1); it is, in fact, Sam's mother, Ivy—a human being, distinguished from the nonhuman when "the animal" (Mills 2018b, 1) clears its throat. What is this strange place, then, in which the narrative opens? The town Mills describes is made up of "broken road" and "dead trees"; a sky that is "old-car blue" (Mills 2018b, 9). These descriptions of place and the natural inherently lend themselves to the human, the dead trees positioned alongside the broken road and the sky described not simply as blue, but as the blue of an old car, a human technology. These integrative depictions of the human world within the nonhuman are revealing; by the time the reader learns of the catastrophe that hits the town—the disappearance of the sea and the decimation of the Australian giant cuttlefish population—Mills has demonstrated that the characters in her novel are embroiled in much more than a simply human problem.

Mills' choice to focus on the Australian giant cuttlefish, or *Sepia apama*, is key here. Considered the largest cuttlefish in the world, these cephalopods group together in their thousands on the rocky reefs of the Upper Spencer Gulf in South Australia where they reproduce (Kassahn et al. 2003, 947). As is the case for most other cephalopods, the Australian giant cuttlefish is well known for being semelparous, or dying following the one single spawning event of their lifetime (Payne et al. 2013, 1). As a non-mammalian marine species that is often defined by its unusual sexual behaviours, the cuttlefish is a provocative choice by Mills as it represents an animal that is in many ways vastly different to the animal that is the human. The reader is forced to reckon with what it is to *be* in the body of an other so different to ourselves.

Donna Haraway's concept of "making kin" bears particular relevance here (2015). Sam's relationship with the cuttlefish is an effort on her part, whether conscious or unconscious, to experience the deep biological connections she shares with this nonhuman—a species found nowhere else in the world but right here, off the shores of Sam's regional town, a place where she grew up and where the cuttlefish in their thousands spend the final moments of their lives. Sam makes kin with this species to force an understanding of what it is that has led humankind down the path of environmental degradation and exploitation. Often, Sam's connection with this species reveals a sophisticated empathy, a connection that Sam is fascinated by as a child but one that makes her physically ill:

> The animal's eyes looked kind and sad. She'd seen them before; she and Ivy had lain on the end of the jetty in the winter with their heads over the edge.… Looking down like that had made her feel seasick. (Mills 2018b, 53)

As this passage shows, illness and feelings of sickness are often mentioned alongside Sam's connection with the more-than-human world, suggesting that to engage in such a connection is a challenging experience because it opposes the worldviews of a (post)industrialised, settler-colonial society that perceives the nonhuman as something to be exploited, destroyed or ignored. At one point, Sam feels "an awful satisfaction [creep] up the back of her neck, like a creature emerging from the mud" (Mills 2018b, 9). She finds it difficult to distinguish between the feelings of the human and those of the nonhuman; the destruction committed against the nonhuman is inadvertently felt by the human, too. Sam's sickness—her diagnosed "dyschronia" (Mills 2018b, 65)—indicates that she cannot exist in both modes of being; that a choice exists to either accept a deeper ecological awareness or accept a post-industrial world in which outdated modes of environmental exploitation, such as coal mining, are still being adhered to despite their unsustainability. *Dyschronia* implies that the two cannot be accepted together. This is evident when Sam attempts to warn others of looming catastrophe in one of her visions: "The eyes looks through her.… Sam wants to wave, yell out, to help somehow, but she can't move" (Mills 2018b, 91).

It is unsurprising, then, that this connection Sam feels to the nonhuman is often unwanted, and her connection the abiotic nonhuman is just as fraught as her connection to the living cuttlefish. Various passages describe a troubled relationship between Sam and the squid sculpture that dominates the town in Sam's future visions, the sculpture alluding to the irreconcilability of industry and ecological well-being. This is emphasised by Sam's uncomfortable bodily reaction to the sculpture—a squid likeness created from the products of the local asphalt industry—suggesting that the human–nonhuman connection is a strange and uncomfortable one in this context: "Its plaster tentacles stretch out, its broken eye is darkly watching. Sam feels a shudder run through her, so deep it might come from the earth under her feet" (Mills 2018b, 89). The idea that this reaction, this feeling of connection to another species, is not contained within her physical body but extends beyond that, to the earth itself, suggests that the human animal, the nonhuman animal and the surrounding natural world suffer the consequences of environmental destruction in equal proportions to each other.

But there is also something more complex occurring here too—a connection that spans the living and the nonliving nonhuman, a connection that is not wholly good or wholly bad. Mills' novel is certainly critical of industry (the ruinous influence of Clapstone's asphalt industry and the strangeness of the giant artificial squid) and post-industrial corporatism (demonstrated via the attempted tourism development scheme), but the abiotic nonhuman sculpture also draws attention to van Dooren and Rose's thoughts on the "nonbiotic" (2016, 87) as equally worthy of consideration in developing more-than-human approaches to ecological problems. The "plaster tentacles" are of both the human and nonhuman, representing the form of the nonhuman animal but created from the product of human industry. This, coupled with a "drilling somewhere below" (Mills 2018b, 90) and the "rattle in her body, from her body, or the earth beneath" (Mills 2018b, 90) that Sam feels when she first sees the squid sculpture, alludes to what Yusoff describes as "geologic forces that compose and differentiate corporeal and collective biopolitical formations" (2013, 780). The rattle is felt simultaneously within Sam's body and within the earth, and so the reader is shown that the human suffers the consequences of environmental destruction just as much as the cuttlefish does. This drilling also represents what was once the livelihood of the town but, at this moment, is controlling Sam's agency, as she is unable to move or act as she watches "as a tentacle drops from the body of the monster" (Mills 2018b, 90). There is the suggestion that the human act of drilling is responsible for the town's ruination, the receding seawaters and the resultant killing of thousands of cuttlefish off Clapstone's shores.

These uncertain distinctions between Sam as a human, the cuttlefish as a nonhuman, and the squid sculpture as nonbiotic suggest that blurring the borders between these modes of existence presents an opportunity to better engage with the nonhuman, bringing forth a more diverse understanding of the more-than-human that goes beyond the living. Sam's distressed reactions to this human–nonhuman connection suggest an uncertainty and strangeness around such a relationship developing in a town that is the product of Western capitalism and industrialisation. Her reactions also challenge the idea of human-ness that Sam has grown up accepting, as well as the notion of "living" in the limited way that it is applied to humans and animals. There is a suggestion that any kind of meaningful connection to the nonhuman is not possible in a world dictated by production and profit, where the dualisms of producer/product, living/nonliving, control who or what is deserving of empathy and attentiveness.

In *The Mushroom at the End of the World*, Anna Lowenhaupt Tsing identifies scalability as key to industrialism, capitalism and, by extension, post-industrialism, leading back to the model of the plantation, which relied upon alienation rather than interconnectedness to ensure productivity (Mintz 1986, cited in Tsing 2021, 40). Tsing writes, "By envisioning more and more of the world through the lens of the plantation, investors devised all kinds of new commodities. Eventually, they posited that everything on earth—and beyond—might be scalable, and thus exchangeable at market values" (2021, 40). This model "congealed as modern economics" (2021, 40), thereby carrying over into post-industrialism. This is exemplified in *Dyschronia* by the corporate speak of the DSC and the tourism development scheme that alienates Clapstone's people from their own home, including the coastal environs and its nonhuman inhabitants that had previously drawn people to this part of the world. The evocation of the town and its surroundings as scaled down to a marketable value—and not considered or appreciated in the context of their broader, relational value—speaks to this sense of (post)industrial alienation that Tsing describes.

As the reader, we are encouraged to wonder whether, after her own experiences detailed later in the novel, Sam is now unable to see the human as purely human—instead, the human is inextricably connected to those animals and the nonliving materials which define our connection to place. If we take Val Plumwood's concept of philosophical animism as described by Rose, Sam's relationship with the cuttlefish and the squid sculpture represents an understanding that "personhood" exists beyond what we know as human (2013, 96). However, this understanding is one that is approached with much difficulty by Sam. The pain, like "a cold steel spike" (Mills 2018b, 118), that results from Sam's ability to see the future of the town and the cuttlefish—a punishment, perhaps, for her connection with a more-than-human personhood—represents what we should fear if Euro-Western instrumentalist approaches to nature and the nonhuman continue unchecked. In Sam's case, the realisation of the human–nonhuman relationship comes too late in a time of environmental crisis. Seeking a more relational connection with the natural world, especially one which many settler-Australians are not familiar with, takes work; Mills' novel suggests that if this is not done soon, then more mistakes will be made on the road to climate catastrophe—potentially "a jumble of infinite possible mistakes", as Sam perceives (Mills 2018b, 354).

In this respect, Mills' novel can be read as posthumanist; Sam's attempt at becoming *with* the nonhuman is reminiscent of those multispecies posthumanists who contend that while we can try and understand the nonhuman other, our attempts are somewhat futile. Maurice Merleau-Ponty's "deep world of untamed perception" (1973, 118) speaks to this. We can attempt to cross this abyss between the human and nonhuman, but it largely remains untamed. This corresponds not just to the nonhuman animal, but the nonliving nonhuman—particularly as humans attempt to grapple with their own role as a geological force in the context of the Anthropocene. Whilst Jacques Derrida contends that it is difficult, if not impossible, to completely disintegrate the limitations within human–animal relationships (2002), Sam's relationship with the cuttlefish and the squid sculpture implies that it is indeed possible, but that the human being that emerges on the other side of this relationship may be vastly different and, in many ways, posthuman. In this context, Sam's transformation into a human with intimate knowledge of the nonhuman reflects the potential of an ecological posthumanist model in increasing attentiveness to more-than-human agency. If we take Haraway's "making kin" argument (2015), Mills is suggesting that if humankind cannot connect with those nonhumans that ultimately define the places that we inhabit, we cannot possibly understand the plight of these beings, let alone ourselves and our relationship with the nonliving materials, such as fossil fuels, that define the Anthropocene.

In terms of understanding nonhumans, marine creatures are almost in a category of their own, especially those which are not mammalian. The form of a cephalopod can be confronting and alien-like to those humans who know little to nothing about it, and yet the Australian giant cuttlefish is so unique to Australia that its extinction would spell the end of another of this continent's endemic animals. In this sense, classically aesthetic values often define whether the general public and those in positions of power empathise with the nonhuman or not. This comes down to the inability of many to view nonhumans in colonised place through a lens that is not purely Euro-centric. The decimation of Clapstone's beach suggests a resetting of these Western notions regarding Australia's native nonhumans. The quintessentially golden beaches and shimmering blue waters of Australia's southern coastline have been transformed by the end of the novel:

> The torn-off pegs of the old jetty were dimly visible in the sand, jagged spikes sticking between jagged rocks, jagged as tombstones. Beyond them, the sea was only a faint blur on the horizon, a graphite stripe between dead sand and sky. (Mills 2018b, 307)

There is a notable point made here regarding the positioning of the nonliving materials of the jetty—the "torn-off pegs" and the "jagged rocks"—in this immense graveyard of ecological carnage. Beyond these materials are the thousands of dead cuttlefish, their rotting corpses covering the exposed seabed. Both play a role in the township's connection to place—the jetty, a location I described earlier as key to Sam's own sense of place and relationship with the sea, and the cuttlefish, the iconic species that previously drew locals and tourists alike to the ocean each year to witness their unique mating event. The abiotic and the biotic nonhuman are inextricably linked here in two ways—firstly, in terms of their role in asserting the settler connection to Clapstone, and secondly, their role in unsettling said connection, when the consequences of the toxic local asphalt industry and the failed development scheme are made known via the ecological devastation of the coast.

Who, though, is responsible for this devastation? Sam as seer bears some individual responsibility, but the collective responsibility of humankind is a key theme of this novel. The unanimous voice of Mills' unknown group of Clapstone townsfolk who narrate some of the novel's chapters and refer to themselves collectively as "we", ask themselves and the reader, "How can we see what we can't imagine?" (Mills 2018b, 13). Through this, Mills depicts the impossibility of addressing environmental destruction if we do not use story and collective action to explore the implications and causes of it. If humans cannot even begin to imagine the consequences of not acting on the destruction of the nonhuman and the natural world in general, how can we imagine life without the nonhumans that make the places we live so unique and the ecosystems healthy enough for humans to survive in?

In her essay "Against Realism", Mills states that her use of the first-person plural, or chorus, in *Dyschronia* represents her "struggle to articulate collective responsibility" (2018a). Even as an individual, Sam—like the chorus—struggles to bring authority to her claims regarding the future; it is therefore not the chorus alone that represents this struggle to take responsibility for failing to act on the decimation of the nonhuman. As the chorus' narrative progresses further into the future from the time of Clapstone's environmental catastrophe, the reader perceives Sam's story moving forwards and backwards in time as she suffers from her illness and a similar inability to understand her responsibility to the town and the cuttlefish. Sam does what van Dooren and Rose describe as "bear[ing] witness" (2016, 89) to the nonhuman others within a multispecies world. However, this act of becoming witness is portrayed as occurring too late—the marine habitat and its nonhuman residents have been decimated, and the human habitat of Clapstone along with it. The only material evidence of the nonhuman that exists in the final chapters of the novel is the abiotic nonhuman in the form of the plaster squid sculpture, perhaps pointing to the industry that created this sculpture as the real cause or, more complicatedly, the difficulty of disentangling the impacts on the human from those on the nonhuman—the human culprit is forced to watch as both they themselves and the nonhuman other are consumed.

The text thereby presents the idea that perhaps settler-Australians are too late to enact a non-Western view of the Australian nonhuman—even if we can develop the ability of multispecies interpretation, the impacts of climate change and environmental exploitation are already being seen. Perhaps such interpretation cannot be left to Sam alone, but rather, as Mills states, the future must be "both individually and collectively" determined (2018a).

This focus on collectivity brings me to an important point about time and its configuration in this novel. Unlike many other works of speculative fiction that present time travel or future knowledge as a means of averting crisis,[1] Mills' text portrays an altogether different paradox: one in which Sam, the seer, perceives a crisis that she cannot avert. Clapstone is described as "slowly being dismantled by gravity" (Mills 2018b, 3), emphasising the inevitable collapse of settler-colonial civilisation in the face of an ever-present, nonhuman power. In this novel, Mills especially emphasises the invisible, inanimate nonhuman of the human world—presences such as gravity and time which humans rely on but which exist whether or not the human does. As the title of the novel and the name of Sam's illness suggests, time is a major theme, threading through the narrative in a way that

[1] For a discussion of time travel and crisis aversion in speculative fiction, see Klotz 2019.

not only highlights Sam's special ability to see the future but emphasises the time experienced by the natural world in stark comparison to the short lives of human beings. Deep time is of particular interest here, not just in how it impacts the human and nonhuman equally—both human and nonhuman beings are ultimately at its mercy—but as a nonhuman force in its own right. How does time function without us, the human *or* nonhuman, experiencing it passing by?

The inclusion of the cuttlefish in this time-focused narrative is significant, given that they live relatively short lives, and their final moments are characterised by a manic breeding frenzy. The disastrous devastation of their population in the novel also appears sudden and unavoidable, whereas the humans who linger behind experience time in a more tedious, less exciting fashion, witnessing the town's livelihood—even as a freakish tourist destination post-catastrophe—gradually diminish:

> Busy seasons come and go, and we adapt to them.... The older demographic likes to pause and wonder. But after a while the older tourists start to dry up as well, soaking away like water into cracked ground. There are days, then weeks, then months when nobody comes... (Mills 2018b, 149)

This juxtaposition of the cuttlefish's devastation with that of the regional human town highlights how time is known differently by different species. Yet, in the context of deep time, the life of an individual human and an individual cuttlefish is almost indistinguishable; a human and nonhuman drop in the ocean of deep time, whilst the nonliving squid sculpture has the potential to outlast both.

Throughout the novel, time in general is often depicted in relation to the nonhuman. It tugs at Sam "like a hook tugs a fish" (Mills 2018b, 91), suggesting that the nonhuman is subject to similar temporal restrictions of the human. This emphasises the urgency of addressing the modern environmental crisis in the context of both human *and* nonhuman mortality, decentring the human whilst still including it. This relates to Serpil Oppermann's discussion of how quantum theory may play a role in challenging the duality of human and nonhuman realities (2008), providing opportunities to analyse the human–nature relationship through an understanding of how the nonbiotic nonhuman—such as time and space—impacts the human and living nonhuman in both similar and different ways.

Just as Mills implements the first-person chorus to demonstrate an inability to acknowledge responsibility throughout the linear part of the narrative, the nonlinear timeline experienced by Sam demonstrates an inability on her part to come to terms with the consequences of her illness for the townspeople of Clapstone and her relationship with her mother. A method employed in other novels of Australian ecofiction, especially those which confront issues of climate change head-on such as Catherine McKinnon's *Storyland* (2017) and Bradley's *Clade* (2015), nonlinear timelines and time-skipping more generally suggest a serious urgency when it comes to problems associated with the modern environmental crisis. This is especially relevant as Australians begin to perceive that the reality of living in a country of raging fires and rising sea levels is fast-approaching. This manipulation of the novel's chronology also depicts a diffusion of responsibility as the narrative slips in and out of different times, the reader suddenly observing old problems being left behind, only to be replaced with new and often more frightening ones as ecocatastrophe progresses.

So, then, does Sam's personal relationship with a nonhuman suggest that the human can properly make kin with the nonhuman, or is such an idea admirable but ultimately impossible? Sam's condition often eliminates her agency, portraying her as a stumbling nomad trying to make sense of the disaster that has befallen her once prosperous coastal town, as well as herself. As Shirm describes, "Sam's inertia begins to look rather like our own when faced with environmental catastrophe" (2018). The reader is drawn to Sam's psychic-like abilities but is also repelled by her inability to take real action on the future that she sees; she and the town of Clapstone are travelling rapidly towards a future that none are prepared for, despite the fact that Sam knows the inevitability of this all along. Mills' message, here, is therefore to act on the issues of climate change, extractivism, habitat destruction, and human disconnection from more-than-human life-worlds using the knowledge we have now before time determines that it is too late.

5.3 The Couch and the Potato Sack: Objects of Place in Yumna Kassab's *Australiana*

Kassab's *Australiana* presents a series of interconnected stories about an unnamed, inland regional Australian town, possibly Tamworth in northern New South Wales (Archer-Lean 2022). Farming has long been the core industry supporting settler residents in this area, with the different stories and characters following a broader narrative through-line of environmental and social upheaval caused by extreme drought and flooding, but also changing attitudes towards and about settler–land relations. Whilst exploring similar themes to Mills' novel, *Australiana* unsettles the traditional singularity or individualism of regional Australian settler narratives via an "interweaving of discrete stories" inspired by Arab storytelling practices, such as those exemplified in *One Thousand and One Nights* (Archer-Lean 2022). As a Lebanese-Australian author with family who are farmers in Lebanon, Kassab is particularly interested in the notion of regionality and the fact that "not many people from the cities spend a lot of time in regional areas" (Chidiac 2024), especially in Australia. Kassab is known for exploring human–land relations and postcolonial cultures and politics across both her fiction and non-fiction, the latter being particularly attentive to cultural geographies and dispossession (see Kassab 2021, 2023) and her fiction often presenting collective rather than singular experiences of place.

Like *Dyschronia*, the novel follows a messy chronology. Unlike Mills' novel, though, there is no single protagonist, although some characters (both living and nonliving) seem to re-occur across some of the interconnected stories, such as a farmer dealing with the fallout of his marriage and the devastating impact of drought on his livelihood, and an elderly woman's couch that becomes embedded in a dry creek after a flood. Farz Edraki identifies "polyphony" as at the centre of Kassab's work (n.d.). As Kassab herself says about her fiction, "what I'm usually writing is the story of a community. There are many viewpoints. Maybe it does arrive somewhere, but maybe it doesn't" (Edraki n.d.).

Most of the human characters are depicted as white, settler descendants attempting to eke out a living in a time when the sustainability of Western-colonial farming practices seems more questionable than ever. What re-occurs more than the human characters are the nonliving objects. There is the aforementioned couch that features prominently across at least three different characters' stories; a brown potato sack in at least two stories and the human dummy that it forms a part of; and lost supermarket trolleys. Some chapters, particularly the later ones, seem almost entirely disconnected from any of the characters or objects that the reader has been introduced to throughout the earlier sections of the book.

The novel's paratextuality references the traditional rural settler novel that *Australiana* is satirising. The front cover depicts a scorching yellow sun and layered red and brown sky over a rural homestead. The title of the novel of course also points to this, the word "Australiana" referring to cultural objects and forms of cultural expression that represent stereotypical white, settler views of the continent. A recognisable example of this would be objects or artworks that depict the infamous bushranger Ned Kelly, viewed in the settler mind as the epitome of the Australian underdog and often portrayed as a working-class "battler" and misunderstood criminal who survived in the Australian bush when he was on the run from British authorities.

There is an exaggerated self-awareness of this amongst some of the novel's characters:

> Enough of this dust bowl his people were determined to call home. Perhaps once this was a country built on the sheep's back but those were the golden years, the gone years... That was a time of perfectly spun wool, the sheep munching happily in the paddock, a rugged pioneer, his shirtsleeves rolled to his elbows, his foot resting on a shovel, eyes challenging you with a dare, eyes that said *I am equal to this land*. (Kassab 2022, 122)

The abiotic in this novel is largely represented by a combination of rural and regional objects and infrastructures associated with drought, floods and agriculture. From furniture to supermarket trolleys, agricultural machinery to sentimental trinkets, roads to fences, the objects and infrastructures of this unnamed inland town are not simply props, but reflect the agency of the abiotic nonhuman in the larger economic, social and ecological machinations of the region.

Like the traditional settler novel, much of the narrative is characterised by concerns about drought and the resulting social issues that pervade small-town life in the absence of water:

> So many dams and so little water in most of them. If there was one silver lining to the drought, it was that drownings were down. Never mind the increase in domestic violence, shootings and suicide. (Kassab 2022, 22)

The flipside of drought is, of course, flood. While I discuss floods and water in more detail in Chap. 7, it is worth considering how Kassab portrays rural concerns about flood through a focus on the abiotic: "We were kicking ourselves for not putting in concrete blocks but the water got above them as well" (Kassab 2022, 49). The most recurrent nonhuman figure in this flood narrative is a couch, originally owned by an older woman whose house is partly cleared out by her family, seemingly against her wishes, only for the house to be swept away in a catastrophic flood:

> They distributed her furniture to charities, they tossed her clothes in the bin, they recycled old letters and photographs. She caught them in the middle of their clean and they promised her they were keeping a few mementos but they never did. If they gave in to her wishes, the house would still be stuffed with possessions she did not need now that she was getting on.... When they heard about the floods in the gully on her property, most of the family came around. From the next hill, they watched Nan's house floating away, and were struck that it had finally happened, the day had finally arrived. (Kassab 2022, 52)

The narrative voice in this particular chapter is a granddaughter of Nan's, who has been eyeing off the couch for herself over many years. Unfortunately, though, the couch finds a new home in a creek bed following the devastation of the flood:

> It now rested on a bed of leaves, sticks and weeds, keeping company with the trolleys, plastic wrappers and empty beer cans—but if she ignored the rubbish, she could imagine the waters dropping, the creek emptying, the trees blooming, the grass springing fresh, and how even out here, the couch wouldn't be a bad place to sit. (Kassab 2022, 57)

The narrative of the couch depicts its journey as a product from Europe making its way to regional Australia—an exemplar of the manufacturing trade relations that are able to bring an expensive designer couch into a seemingly unremarkable Australian regional town: "The couch had travelled from a European country, it had been loaded on a boat, it had travelled up the highway, it had been handled by five strong men" (Kassab 2022, 56).

The couch intermingles with the living nonhuman in the creek bed, providing an opportunity for the human and nonhuman to connect in a time of social and environmental instability for the town:

> A designer couch sitting in the mud ... it was the perfect spot for a coffee, to forget that humans existed, that the drought had receded but was not yet over, that their town had gone to level four restrictions, that the greenery was just a veneer. Dig a heel in the ground and the dust flies up and the hard days are not yet done, not by a mile..." (Kassab 2022, 61)

This story of the couch is attentive to the relationship between broad-scale, global production and the value of such items at the regional, local level in small-town Australia. This returns me to Tsing's point about the importance of scalability in (post)industrial capitalism. When objects are viewed only in terms of their scalable production value—as commodities that are produced via the "smooth expansion" of global manufacturing rather than the "indeterminacies of encounter" (Tsing 2021, 38) that upset the progress narrative of Western industrialism—then it is difficult to see them as anything by an extension of the global post-industrialist, capitalist ideologies that created them. It is only when such objects are removed or affected by the more-than-human life-worlds around them, such as through floods or droughts, that their commodified potential is drastically altered or disappeared. Many of the objects described in Kassab's novel may be the products of scalable production in the global setting but, at the regional level, we see how their value and purpose is altered in a way that actively goes against global capitalist ideologies. This is particularly true of the couch which is assigned value in a more relational context (as a place for people to spend time in nature but also in terms of how its new position in the creek bed draws attention to the agency of flood and water)—a very different kind of value from its commodified desirability as "the most expensive item in [Nan's] home" (Kassab 2022, 56) with its own "gravitational pull" (57).

A short chapter titled "Assemblage" points towards the complex relationality between objects and the human, describing the construction of a "dummy", its purpose somewhat ambiguous to the reader, although it is implied in the following chapter that it is to be used for therapy purposes by the woman who makes it: "a punching bag, it could be shot at, there was the therapy of making it, the thread and needle for a mend" (Kassab 2022, 99). In the initial chapter describing its assemblage, though, its purpose and maker are conspicuously absent. Beginning with "a potato sack. Brown, rough" (Kassab 2022, 97), the chapter reads as an instruction manual for the dummy's assemblage. "Tie a rope around its centre. This is the waist", it continues, "for the legs, pyjama pants—the colourful sort—will do the job" (Kassab 2022, 97). Each material that is subsequently added to the dummy represents a new body part: "the head: a football", "wool for hair", and "for the arms, try the snakes that keep a draught from a warm room" (Kassab 2022, 97–98). The chapter ends with the assertion that "a dummy's purpose is to be used" (Kassab 2022, 98). Connections between the living and the abiotic are further solidified in a later, absurdist chapter describing the town being overrun by animals who take over via their use of human-made objects: "The kangaroos tore in on their scooters and skateboards, the echidnas carried guns, the koalas tore out the electrical wiring of the town" (Kassab 2022, 189).

These links—between the dummy and the human, the objects and the animals—further reflect the complex entanglements between life (and nonlife) worlds that exist beyond the confined binaries of producer/product, subject/object that define capitalist production. The animals' use of commodities destabilises their own status as living products in the anthropocentric system of Western capitalism—as meat, as fur, as tourist drawcards—as they overpower the town via the use of these nonliving commodities. Similarly, the dummy is assembled using various individual objects, with the intention of representing the form of a human by the end of the process. The dummy is a collective assemblage of these various objects, and it is only through such an assemblage that an individual image of the human takes form. These instances in the novel challenge the alienation of global, post-industrial production and consumerism, suggesting more of an attunement to "diverse modes of engagement, entanglement, and cobecoming" (van Dooren and Rose 2016, 80) that trouble any distinct lines between human and nonhuman, living and nonliving.

As in *Dyschronia*, Kassab's novel explores the role of sculptures and statues in facilitating settler narratives of progress, highlighting the irony of this in the face of economic and social upheaval in Australian rural and regional communities in the twenty-first century. The final chapters of the novel are concerned specifically with legends of bushrangers, especially the ways in which they are memorialised in white settler culture. These chapters focus on two bushrangers in particular—Captain Thunderbolt, or Frederick Wordsworth Ward, and Ned Kelly, both real historical figures. Kelly, though, is far more famous, Kassab making this point by not mentioning him specifically by name with the assumption that the reader is familiar with him: "You know his name, the other one. He is an icon, his mask is known the country over" (2022, 277).

Kelly's infamy is due in large part to his iconic armour, made from the iron of old farming machinery, the helmet of which is one of the most widely identifiable symbols of settler-colonial Australia—the essence of "Australiana". As Kassab describes, "People have him tattooed on their biceps, their calves, his black mask, like a letterbox, winking with a glisten from sweaty skin" (2022, 277). Artistic recreations of Kelly in his armour exist in many contexts across Australia, including several statues, such as the "larger than life" (Kassab 2022, 277) one located in Glenrowan where he famously took his final stand against police.

Kassab's focus, though, is primarily on the lesser-known Captain Thunderbolt, although he too has been memorialised with a "bronze statue" that "is whispered to move" (2022, 275) and is graffitied in the novel by parties unknown, perhaps in protest of the enduring colonial powers in contemporary Australia. The comparison between Thunderbolt and Kelly draws attention to the role of objects in settler-Australian mythologies of place—the claiming of land and belonging through material memorialisations of these icons of colonial resilience and strength, despite (or because of) their status as criminals.

At first glance, these statues do not explicitly reflect the true violence of colonial conquest nor the environmental devastation caused by colonial extractivist practices and the development of town infrastructures, both of which are required for the statues to exist in the first place. But just as Kelly's armour, fashioned from old ploughs, speaks to the legacies of regional agriculture and production in colonial Australia, the metals used to make the statues reflect the extractivist cultures that were required to obtain these materials. Such materials are therefore used in the name of colonial progress—in farming, infrastructure, and otherwise, but also in the formation of colonial symbols, such as Ned Kelly's helmet and the statue that memorialises both the man and the armour.

It is difficult to ignore the synergies between Kassab's interrogation of the settler-colonial statue and Mills' use of the squid sculpture in celebrating regional economic flourishing at the same time that such flourishing is rapidly extinguished by ecological devastation. In the novel's penultimate chapter, Kassab writes from the perspective of Thunderbolt himself as he experiences the moment he is shot and killed by police at Kentucky Creek in northern New South Wales. The description mirrors the floods described earlier in the novel and the couch in the creek: "waters murky with suffocation … my end will be a pond … it will be a creek … my horse shot" (Kassab 2022, 288). The linkage made here reveals the interconnections between the human and nonhuman across time that result in this myth and memorialisation of Thunderbolt—the man, his horse, the creek and, later, his statue, and the rock, "a towering grey monolith", that marks his grave (Kassab 2022, 285).

This use of the abiotic nonhuman in such circumstances is problematic—the statues cover up the crimes of dispossession, reinforcing a dichotomy of human/nature, town/bush, coloniser/colonised as the solidity of the metal bushranger lookalike attempts to assert settler dominance and belonging. When an author like Kassab digs a little deeper, though, the intricacies of human/land relations—even those involving the settler—are revealed to be deeply indebted to and, at times, upended by the nonhuman, even if this is not recognised in the settler mythos itself. Passages such as the above and others in Kassab's novel reflect this nuance, unsettling white settler claims to land while simultaneously revealing the complexity of settler–land relations in a regional Australian setting under threat of drought and flood.

5.4 Conclusion

Dyschronia concludes with Sam perceiving the universe and its "jumble of infinite possible mistakes" (Mills 2018b, 354), emphasising the notion that to comprehend nonhuman ways of being is to accept that on some level, the Euro-Western, settler-colonial way of being will always be problematic and essentially counterproductive in the face of climate change; it will perpetually oppose those complex, meaningful human–nonhuman relationships that may be a solution to the current environmental crisis.

Australiana achieves something similar through its disconnected narrative chronology and connected nonliving characters, the regional "everytown" setting of the novel and the depiction of stereotypical rural/regional objects, such as farming equipment and infrastructures, drawing attention to the impact of post-industrial shifts in regional Australia, and what this means for human–nonhuman relations in times of extreme drought and flood. As Kassab herself writes in her essay "National Accounts", "everywhere are your references to the empire claiming the land" (2021, 2).

What both novels show is that the nonliving nonhuman is not to be disregarded, in either its literary or material form. Across different contexts, it has the potential to function as either a powerful endorser or opponent of settler-colonial ideologies. Understanding the objects, infrastructures and other abiotic aspects of the more-than-human, such as deep time, is therefore essential to understanding the human and the living nonhuman and challenging the dualisms that separate the categories of living and nonliving from one another.

References

Archer-Lean, Clare. 2022. Colonial and nationalist myths are recast in Yumna Kassab's *Australiana*. *The Conversation*, April 13. https://theconversation.com/colonial-and-nationalist-myths-are-recast-in-yumna-kassabs-australiana-178881. Accessed 7 Feb 2025.

Bell, Daniel. 1973. *The Coming of Post-Industrial Society: A Venture in Social Forecasting*. New York: Basic Books.

Bradley, James. 2015. *Clade*. Melbourne: Hamish Hamilton.

Bradley, James. 2018. James Bradley Reviews *Dyschronia* by Jennifer Mills. *Australian Book Review* 399.

Chidiac, Valerie. 2024. Yumna Kassab on teaching, writing, and the Parramatta of it all. *Honi Soit*, May 15. https://honisoit.com/2024/05/yumna-kassab-on-teaching-writing-and-the-parramatta-of-it-all/. Accessed 7 Feb 2025.

Derrida, Jacques. 2002. The Animal That Therefore I Am (More to Follow). *Critical Inquiry* 28 (2): 369–418.

Edraki, Farz. n.d.. Parramatta Profiles: A conversation with novelist Yumna Kassab. *Powerhouse Museum*. https://powerhouse.com.au/stories/writing-life-a-conversation-with-novelist-yumna-kassab. Accessed 7 Feb 2025.

Haraway, Donna. 2015. Anthropocene, Capitalocene, Plantationocene, Chthulucene: Making Kin. *Environmental Humanities* 6 (1): 159–165.

Henningsgaard, Per. 2007. Regional Literature and "The Liminal": Exploring the Spaces In-Between National and International Literatures. *Limina* 13 (3): 12–19.

Henningsgaard, Per. 2010. A Postcolonial Perspective on Regional Literature in Australia. In *Postcolonial Issues in Australian Literature*, ed. Nathanael O'Reilly, 57–74. Amherst: Cambria Press.

Hughes-D'Aeth, Tony. 2023. Australian Fiction in the Anthropocene. In *The Cambridge Companion to the Australian Novel*, ed. Nicholas Birns and Louis Klee, 289–304. Cambridge: Cambridge University Press.

Kassab, Yumna. 2021. National Accounts: The Conquest of Land and Dream. *Meanjin* 80 (3): 1–4.

Kassab, Yumna. 2022. *Australiana*. Ultimo, NSW: Ultimo Press.

Kassab, Yumna. 2023. The Rooster and the Watermelon: Yumna Kassab on the potency of symbols. *Sydney Review of Books*, December 17. https://sydneyreviewofbooks.com/essays/the-rooster-and-the-watermelon. Accessed 7 Feb 2025.

Kassahn, Karin S., Stephen C. Donnellan, Anthony J. Fowler, Karina C. Hall, Mark Adams, and Paul W. Shaw. 2003. Molecular and Morphological Analyses of the Cuttlefish *Sepia apama* Indicate a Complex Population Structure. *Marine Biology* 143 (5): 947–962.

Klotz, Marcia. 2019. Of Time Loops and Derivatives: Christopher Nolan's *Interstellar* and the Logic of the Futures Market. *CR: The New Centennial Review* 19 (1): 277–297.

LeMenager, Stephanie. 2017. Climate Change and the Struggle for Genre. In *Anthropocene Reading: Literary History in Geologic Times*, ed. Tobias Menely and Jesse Oak Taylor, 220–238. University Park: Pennsylvania State University Press.

McKinnon, Catherine. 2017. *Storyland*. Sydney: HarperCollins Publishers.

Merleau-Ponty, Maurice. 1973. *The Visible and the Invisible*. Evanston: Northwestern University.

Mills, Jennifer. 2018a. Against Realism. *Overland Literary Journal* 233.

Mills, Jennifer. 2018b. *Dyschronia*. Sydney: Picador.

Mintz, Sidney. 1986. *Sweetness and Power: The Place of Sugar in Modern History*. Harmondsworth, UK: Penguin.

Moretti, Franco. 1998. *Atlas of the European Novel, 1800–1900*. New York: Verso.

Oppermann, Serpil. 2008. Seeking Environmental Awareness in Postmodern Fictions. *Critique: Studies in Contemporary Fiction* 49 (3): 245–246.

Payne, Nicholas L., Edward P. Snelling, Jayson M. Semmens, and Bronwyn M. Gillanders. 2013. Mechanisms of Population Structuring in Giant Australian Cuttlefish *Sepia apama*. *PLoS One* 8 (3): 1–7.

Rose, Deborah Bird. 2013. Val Plumwood's Philosophical Animism: Attentive Interactions in the Sentient World. *Environmental Humanities* 3 (1): 93–109.

Shirm, Gretchen. 2018. *Dyschronia* Review: Jennifer Mills' Dark Vision of an Australian Future. *The Sydney Morning Herald*, February 15.

Smyth, Elizabeth A. 2022. Writing an Australian Farm Novel: Connecting Regions via Magic Realism. *Text* 26 (2): 1–16.

Steer, Philip. 2021. The Climates of the Victorian Novel: Seasonality, Weather, and Regional Fiction in Britain and Australia. *PMLA* 136 (3): 370–385.

Tsing, Anna Lowenhaupt. 2021 [2015]. *The Mushroom at the End of the World: On the Possibility of Life in Capitalist Ruins*. Princeton: Princeton University Press.

Van Dooren, Thom, and Deborah Bird Rose. 2016. Lively Ethography: Storying Animist Worlds. *Environmental Humanities* 8 (1): 77–94.

Wright, Alexis. 2014. *The Swan Book*, B Format Edition. Artarmon: Giramondo Publishing Company.

Yusoff, Kathryn. 2013. Geologic Life: Prehistory, Climate, Futures in the Anthropocene. *Environment and Planning D: Society and Space* 31 (5): 779–795.

CHAPTER 6

Bushfire, Drought and Settler-Colonial Culpability in the Eco-crime Novel

6.1 Settler-Colonial Eco-crime and Its Critics

In 1997, Stephen Knight described Australian crime fiction as a genre that is "thriving but unnoticed" (1). While in recent years the genre has gained more attention amongst both academics and reviewers, it is still under-considered in ecocritical discussions of Australian fiction. This is despite many works of Australian crime fiction that reflect important ideas about the way Australians—in particular, non-Indigenous, settler Australians—understand and interact with the environment and the more-than-human.

This chapter discusses how modern Australian crime fiction often portrays the troubling relationship between human violence and the treatment of the nonhuman, with a particular focus on drought and bushfire in Jane Harper's *The Dry* (2016) and Chris Hammer's *Scrublands* (2018). Such a relationship indirectly alludes to the impact of a changing climate on Australian communities and ecosystems, suggesting that popular genre fiction can contribute in significant ways to broader environmental considerations. Such texts reveal to readers the criminal nature of anthropogenic climate change and the settler-colonial destruction of Australian habitats.

Whilst place is a prominent element of crime fiction, playing the role of the scene in which crimes and their investigations take place, there is just as much to be said of the significance of nonhuman habitat—even if

R. Fetherston, *Theorising the Postcolonial Eco-Novel*, Literatures, Cultures, and the Environment,
https://doi.org/10.1007/978-3-032-04466-2_6

significantly altered through colonisation—in Australian crime fiction and what it reveals about the relationship between ecological destruction and Australia's violent colonial past. As discussed earlier, I do not consider habitat as a replacement term for place, but rather as one element of a text's broader sense of place and, as argued in Chap. 4, an agential force in its own right.

As described by Sofia Sjö (2016), the strong associations between crime fiction and popular culture worldwide have meant that this genre is often viewed as simple entertainment. Scholars, however, are now more clearly demonstrating the influence of crime fiction on modern culture more broadly, with more researchers choosing to focus on this genre in their work (Sjö 2016, 1). It is only in very recent years that the ecocritical value of crime fiction has been noticed, as evidenced by a crime-focused issue of *Green Letters* edited Jo Lindsay Walton and Samantha Walton in 2018, Stewart King's work on "crimate fiction" (2021) and Nathan Ashman's *Routledge Handbook of Crime Fiction and Ecology* (2023). The fact that crime fiction often attracts a broader audience provides an opportunity to encourage social change amongst a wider variety of readers, and yet it is for this same reason that it has arguably been overlooked by scholars in past decades. As Knight argues, "it would be a crime not to investigate it" because of how strongly crime fiction speaks, through its popularity and ephemerality, about "the concerns, beliefs, fears and obsessions of people" (1997, ix), including those associated with ecological crises.

Like other popular fiction genres, Australian crime fiction began in the pages of newspapers as serialised stories. From the 1850s onwards, Melbourne's *Leader* was one colonial newspaper that published sensational tales of adventure and romance. The latter half of the nineteenth century and the early twentieth century saw the publication of previously serialised works as books, as well as reprints of older colonial stories. In their summary of what they call "these rekindlings of early popular works", Ken Gelder and Rachael Weaver argue for the significance of the crime genre within the colonial scene (2008, 2). In fact, "Australia's first novel" (Gelder and Weaver 2008, 2)—*Quintus Servinton* (1830) by Henry Savery—was a crime novel. Fergus Hume's *The Mystery of a Hansom Cab* (1886) is a work of colonial Australian crime fiction that went international, selling 20,000 copies in Melbourne (the setting of the narrative) and half a million copies globally (Gelder and Weaver 2008, 2). Significantly, Hume's novel has never been out of print (Sussex 2015, 7), was republished as a Text Classics edition in 1999, and has seen renewed interest this century with a television adaptation released in 2012 (Seet 2012; Louttit 2015, 59).

It therefore seems essential to understand how the roots of Australian crime fiction stem from early colonial literature if we are to understand the ways in which the genre today encapsulates themes of colonial violence through the lens of environmental exploitation. Interestingly, Gelder and Weaver describe how, in many colonial crime stories, "the Australian bush ... can even seem to bring about the kinds of degradation and debasement that enable criminal impulses to flourish" (2008, 6). This portrays the natural world as an exacerbator of human violence, reducing the culpability of those humans responsible. Such a viewpoint is related to the large-scale degradation of Australia's environments since the time of early European colonisation, during which white settlers could become utterly sullied by "the lawlessness and corruption of the frontier" (Gelder and Weaver 2008, 6). The destruction of Australia's native habitats and species has, of course, been linked to the settler-colonial desire for "dominion of civilisation over nature" (Langton 1996, 16). Early colonial stories, especially those involving crime, often encouraged this perspective, relating to what Gelder and Weaver describe as "the other side of itinerant life [for colonisers]—the drive towards colonial settlement and stability" (2008, 3). This dominion over nature could, in other words, be referred to as the logic of the colonial project to police and dominate the Australian natural world in order to establish, and profit from, the colony.

Before I continue in this vein, though, it is important that I define my use of the term "nonhuman" in this chapter. In both *The Dry* and *Scrublands,* what I term the "individual nonhuman" is sometimes represented by native animals and plants. However, it is more often the plight of the individual non-native nonhuman that is emphasised in these narratives—those domestic animals associated with settler-colonial agriculture, as well as wild non-native species, such as the European rabbit, that traverse these agricultural landscapes.

Additionally, and perhaps most significantly, there are also nonhuman forces that play a central role in these texts, such as heatwaves, drought and bushfire, all pertaining to Australia's broader climate. I refer generally to these nonliving, or abiotic, nonhumans as "nonhuman phenomena". Whilst evidence shows that some of these "natural" phenomena, as they are often termed, are, in fact, largely influenced by anthropogenic climate change (such as intense drought and bushfire) (Jones et al. 2013), these events still exist externally to human societies and cultures, impacting both human and nonhuman lives. For the purposes of this chapter, then, I consider these phenomena as nonhuman, but through a lens that perceives them as forces that are, and will continue to be, exacerbated by anthropogenic climate change.

In regard to crime fiction, Walton and Walton contend that the "generic features [of crime fiction] offer opportunities to reflect on the forms and functions of environmental criticism and ecological narratives more generally" (2018, 3). It is also significant that in his reflection on Tim Flannery's ecological history of Australia, *The Future Eaters* (1994), David Farrier argues that Flannery notes "a connection between historical/colonial and future/ecological acts of violence"—crimes that are often "without a body" (2018, 876). Discussions of colonial violence and the associated environmental devastation are replete with intimations of crimes that are committed not only against fellow humans but against the nonhuman as well. While Farrier contends that "in the Anthropocene there is no single body, culprit or scene by which to definitively identify global ecological crimes" (2018, 878), the links made between settler-colonialism and ecological crisis in the novels of Harper and Hammer suggest that while we might not have a body, the finger can indeed be pointed at settler-colonial domination as one party responsible for intensified environmental conditions in Australia, and worldwide. In this sense, crime fiction provides a lens through which readers may better understand colonial crimes.

Significantly, many contemporary works of Australian settler writing conceptualise nature as something to be controlled by the coloniser. The horrific depiction of kangaroo-shooting in Kenneth Cook's *Wake in Fright* (1961) is one notable example and, more recently, the characters of *In the Winter Dark* (1988), one of Tim Winton's early novels, view the bush "as either romantic or demonic" and are "unable to conceive of alternative ways to relate to their surroundings" (Cordier 2018, 58). Problematic colonial perspectives on nature and the nonhuman still find a voice in Australian literature today, and it is key that both writers and literary scholars continue to problematise and challenge these notions in their work, with particular attention paid to genre works, such as crime fiction, that may reach a broader audience.

Knight claims that crime fiction—"that massively popular genre that is co-extensive with the expansive and appropriative societies of the nineteenth and twentieth centuries"—often represents the ways in which colonised land is picturesque, but also dangerous. In Australia, he claims, popular culture portrays the continent as a place that is "both dramatically beautiful and life-threatening, as if the merits and risks of imperial possession remain a part of the national consciousness" (2006, 17). These clear links between land and violence raise interesting questions about the role that Australian crime fiction plays in broader ecological and place discourse in Australia. Like *Scrublands*, many other Australian crime novels portray strong ties between crime and place in their titles, such as Mark Brandi's *Wimmera* (2017), Candice Fox's *Crimson Lake* (2017) and Emma Viskic's *Resurrection Bay* (2015). Generally, location factors significantly into crime fiction narratives (Knight 1997, 143), and so it perhaps goes without saying that the bush and other Australian natural places often stand at the forefront of many works of Australian crime. Nevertheless, I argue that such texts also form a part of the discourse regarding Australian environmental perspectives and understandings of nature and not simply place, particularly given that a focus on place, as a human-centric concept, does not necessarily go hand in hand with environmental considerations.

According to Knight, Australian crime fiction has over the years addressed a range of typically "dissenting positions", such as gay rights, racial tensions and pro-criminal perspectives. He also states that crime fiction more generally may also be considered "an early, often the first, voice to respond to new social and cultural encounters generated by the colonial situation" (2006, 25). It is perhaps no surprise, then, that environmental catastrophe has factored into the genre in significant ways this past decade, and that "the colonial situation" is also a key theme. Furthermore, Michael Pollak and Margaret MacNabb contend that crime fiction has long demonstrated a social conscience and that in the latter decades of the twentieth century, this conscience has included environmental issues (2000, 99).

Comparatively, Gelder comments on how Australian literature is often said to be "politically progressive", whilst popular fiction—crime fiction, in this case—is "conservative, even reactionary" (2001, 113). How, then, can one understand crime fiction as a potentially resistant form of popular fiction through its depiction of the nonhuman and of ecological catastrophe? While *The Dry* and *Scrublands* do not necessarily venture too far from the formulaic nature of popular crime fiction, I consider these texts unique in their ability to utilise this formula to engage with current ecological crises in a specifically Australian, rural context—even if indirectly. It is perhaps due to their status as popular genre texts that these novels are well-positioned to engage some readers in the concerns and realities of environmental crises, whether these be specific events such as Australia's 2017–2019 drought and the 2019–2020 Australian bushfires or broader, albeit related, concerns such as the overarching climate crisis.

It is notable, too, that noir tropes play a role in climate fiction internationally, such as in Paolo Bacigalupi's *The Windup Girl* (2009) and *The Water Knife* (2015), Liu Cixin's *The Three-Body Problem* (2014) and David Mitchell's *The Bone Clocks* (2014). These novels, amongst others, engage with themes of crime and mystery in their exploration of the climate crisis. In particular, one of the main themes of these novels is culpability—who is responsible for anthropogenic climate change, and how can justice be served? This idea is discussed by Sarah Dimick in her analysis of Antti Tuomainen's *The Healer* (2013), a work of Finnish "climate crime" fiction. Specifically, Dimick asks:

> In a time when we, as humans, are collectively but disproportionately capable of altering the atmosphere, how do we reconceptualize culpability? How do we understand concepts like intentionality or guilt when agency is distributed across a species instead of contained within a single individual? (2018, 20)

This emphasis on guilt and culpability is shared by many climate fiction texts worldwide, and so it is not surprising that crime fiction is also well placed to explore conceptualisations of both individual and collective crimes in the context of environmental catastrophe.

In comparison to climate fiction, works of crime fiction like *Scrublands* and *The Dry* may, in fact, have more potential to ignite an awareness of the Anthropocene through a less obviously didactic telling of the impacts of climate change. Timothy Morton's understanding of how we come to a moment of ecological awareness says much about this:

> The darkness of ecological awareness is the darkness of noir, which is a strange loop: the detective is a criminal. In a strong version of noir the narrator is implicated in the story: two levels that normally don't cross, that some believe structurally can't cross. We 'civilised' people … are the narrators of our own destiny. Ecological awareness is that moment at which these narrators find out that they are the tragic criminal. (2016, 9)

Like a crime reader who discovers that the first-person narrator of a novel has been the murderer all along, the moment of ecological awareness takes the form of a dark epiphany. The texts discussed in this chapter can be read as an intersection of these two constructs—the unreliable crime narrator, and humans as the unreliable narrators of their own destinies, implicated in the ecological crimes occurring around them. The implication of the narrator in the crime is relevant to Kate Rigby's discussion of "ecoprophetic witnessing" in which the prophet—or, in the context of my research, the narrator or detective of a crime novel—"is both implicated in and wounded by the wrongdoing that is shown to be driving his or her world headlong into catastrophe" (2009, 178). As in many works of crime fiction, the narrator is similar to the prophetic voice in that they usually do not narrate "from a place of purity" (Rigby 2009, 178). They are more often than not implicated, or become implicated, in the crimes they are investigating. So, too, is the ecoprophetic witness both investigating and involved in the ecological crises to which they are privy.[1]

Deborah Bird Rose expands on this in her work on "Anthropocene noir", stating that

> … we are spectators in the unmaking of the world we have known; we are spectators in the mass deaths of other creatures and in the misery of numerous and diverse forms of life including humans; we may indeed become spectators of our own demise. (2013, 211)

The detective is a spectator in this unmaking of both humans and nonhumans, but, as according to Rigby, they may also be involved, whether knowingly or unknowingly, in the crimes they are attempting to solve. Relatedly, Walton and Walton insist that "in an era of widespread environmental crisis, the detective's reassuring and restorative functions must … be reconsidered" (2018, 2). The detective may therefore be simultaneously a spectator, a criminal and a restorer of justice in the context of the climate emergency.

What role, then, does the nonhuman play in such realisations? Significantly, Knight highlights the connection between what he refers to as "zero-setting" stories (Australian crime stories set in a non-discernible geographical location) (2006, 19) and the land's ability to enact vengeance on the criminal:

> Bush fires are the favourite force of vengeance … closely followed by dangerous snakes and floods as the agents of justice. It is a curious displacement

[1] I have argued something similar in a discussion of Briohny Doyle's *The Island Will Sink* (2013) (Fetherston 2019).

> of human agency onto the land, as if it has itself become the law, as if having declared terra nullius has given that terra a curious quasi-human power of its own. Are the colonisers the criminals that the stolen land will ironically avenge? (20)

Whilst the texts examined in this chapter are not strictly "zero-setting" crime narratives, they do exhibit this trope of natural justice or revenge. Are the crimes depicted by Harper and Hammer therefore in some way symbolic of the crime(s) of European colonisation, including those committed against the nonhuman and natural world? Is this natural form of justice representative of nonhuman vengeance for the anthropogenic climate emergency and other environmental devastation? These questions will become relevant as I discuss the representation of the nonhuman in the selected Australian crime fiction texts.

Interestingly, Knight also draws attention to a possible link between the continuing decimation of native Australian habitats—"the bush", as he terms it—and a change in the way that Australian crime fiction is being recognised globally. Knight argues that in the past, those writers who have utilised the "bush myth" for public consumption within crime fiction have been successful overseas (2011, 72). Now, that paradigm is shifting as the "bush myth" becomes more and more displaced by Australia's overexploitation of natural resources and the impacts of climate change. As will be discussed, Hammer's and Harper's works engage with some quintessential settler-colonial ideas regarding the bush (which perhaps explains the favourable response to *The Dry* worldwide[2]), but they do so through the lens of drought as an ecological catastrophe. This is perhaps an adaptation of the bush myth for Anthropocene times, indicating an awareness that Australia's natural world, and the way settler-colonisers interact with it, is changing.

Knight argues that nineteenth-century Australian crime fiction played a major role in early understandings of settler-colonial identity (2006, 22). A developing postcolonial self-consciousness in later years resulted in a rejection of police authority in crime narratives. This can be strongly linked to broader anti-authoritarian sentiments still prevalent in Australia today that position criminals, like the infamous bushranger Ned Kelly, as dark heroes attempting to survive within suffocating colonial systems of

[2] Of particular note is a review of *The Dry* published in the *New York Times* in 2017 (Maslin).

domination. The modern relevance of anti-authoritarianism within postcolonial self-consciousness—such as the enduring popularity of the Ned Kelly myth and more contemporary narratives that glorify modern Australian criminals, such as the television show *Underbelly* (Haddrick and Pam 2008–2013)—alludes to what Bill Ashcroft et al. say about the postcolonial: it is a term which "cover[s] all the culture affected by the imperial process from the moment of colonization to the present day" (2002, 2). In other words, the postcolonial is not free from colonial systems and processes and the perpetuated myths and sentiments regarding the criminal; anti-authoritarianism is still rife within Australian settler culture, carrying over from the colonial time to the postcolonial. As will be discussed in relation to *Scrublands*, there was also the rise of what Knight describes as "zero-policing crime stories, where the just outcome is achieved without a detective and where the moral resolution is often some way natural, as when execution occurs through fire or flood" (2006, 24). The history of Australian crime fiction, then, seems closely linked with developing colonial and postcolonial identities as well as the natural environments in which such identities were formed.

The type of Australian crime fiction discussed in this chapter plays a role in "the construction of national culture and self-consciousness", as described by Knight (1997, 3). When I refer to the genre, I refer to it with the assumption that most if not all texts within it address "in some way or other … this country and its issues" (Knight 1997, 3). I also understand it as sitting within the broader genre of Australian Gothic, which Gina Wisker describes as being "replete with hidden, misrepresented and misunderstood histories and a consistent concern with guilt, identity, contradictions and confusions" (2021, 384). As I will go on to explore, contemporary Australian crime novels speak strongly to such themes, particularly in the context of settler-colonial culpability. Whilst some Australian crime writers produce novels that are set outside of Australia,[3] the genre I discuss in this chapter does not simply include all crime novels written by Australians, but rather those that perpetuate Australian settler-colonial cultural discourse in some way through their Australian setting.

A key point here is the intimate connection between Australia's ecological crises and the crimes of genocide and dispossession that have been committed against Indigenous Australians. Ross Gibson (2002) explores

[3] For example, Australian author Christian White's bestselling debut novel *The Nowhere Child* (2018) is primarily set in Kentucky, USA.

the insufficiency felt by Europeans who attempted, and often failed, to tame wild tracts of nature in so-called new worlds; these areas are referred to as "badlands". He states that "tales of murder and itinerancy in wild country are as old as the story of Cain in the killing-fields of Eden" (2002, 14). The Australian badlands, with their history of settler violence enacted against Indigenous peoples and the violent destruction of the land itself, can be likened to an "immense, historical crime-scene" (Gibson 2002, 1). Emily Potter argues that failing to acknowledge and engage with these violent histories of place upholds Australia's colonial national identity and subsequently continues the destructive colonial practices which wreak havoc on marginalised Indigenous communities and the natural world (2014, 82).

Drawing on these concepts, I relate the atrocities portrayed in the selected Australian crime novels to the European invasion of Indigenous Country and the subsequent violence committed against both people and Country which continues today, often in the name of so-called agricultural progress. As Tony Birch (2017) contends, "the usurpation of land not suited for wide-acre agricultural farming had led not only to the appropriation of Indigenous land, but also the destruction of local ecologies and the wasteful use of natural resources such as water and soil". Knight contends that crime fiction is well placed to comment on both "crimes domestic and crimes colonial" (2006, 33). Australian crime fiction is, then, a means by which readers are exposed to ideas of colonial violence in literal and symbolic ways. Even when crimes are resolved in these fictional narratives, violent acts perpetuate a lack of resolution; the protagonist may have solved the crime, but the broader issue of colonial violence is interminable.

6.2 Drought and Unsettled Belonging in Jane Harper's *The Dry*

The Dry offers readers a protagonist who, as described above, is a spectator and a key player in the events of the narrative: a police authority figure who is implicated in a crime that took place in his hometown when he was a teenager, which he is now attempting to solve. *The Dry* is Harper's debut novel, winning the 2015 Victorian Premier's Unpublished Manuscript Award prior to its publication in 2016. It has also been adapted by Robert Connolly into a film starring Eric Bana, which proved incredibly popular with audiences on its release in 2020. Since her debut novel, Harper has released two sequels to *The Dry*, titled *Force of Nature* (2017b) and *Exiles* (2022). So far, *The Dry* remains largely underexplored by scholars, although in their work on perspectives of rural Australian cultural sustainability, Catherine Driscoll et al. contend that Harper's novel represents "the oppositional relationship between the Australian city and its non-metropolitan hinterland" (2017, 1). Sue Turnbull has also investigated the novel's depiction of the relationship between ecological crisis and domestic abuse in rural communities (2023). The narrative of *The Dry* centres on Aaron Falk, an investigator for the Australian Federal Police who has taken a leave of absence from his work in Melbourne to attend the funeral of a childhood friend, Luke Hadler, in Falk's rural Victorian hometown of Kiewarra. Luke is suspected of committing murder-suicide, killing his wife and son and then himself, supposedly due to the pressures of a devastating drought that has impacted his farm and the broader township. As indicated by the novel's title, drought and heat are a constant presence throughout the narrative, highlighting the plight of farmers and townsfolk whose livelihoods are under threat. However, Falk is unconvinced of Luke's guilt and spends his time in Kiewarra trying to uncover the truth, as well as reflecting on a secret he has kept since his school days regarding the murder of a childhood friend, Ellie Deacon.

Unlike the protagonist of Hammer's *Scrublands*, the reader may see a more quintessential crime hero in *The Dry*'s Aaron Falk due to his status as a police investigator. This is not a "zero-policing crime story" as Knight describes (2006, 24), although it is perhaps notable that Falk solves the above cases whilst on leave from the Federal Police and that he does so in the very personal setting of his hometown. Both murders are closely linked to Falk's past and he arguably solves the crimes not as a detective but as a past resident of Kiewarra and is therefore heavily implicated in its histories and its present.

Some of the main representations of the nonhuman in *The Dry* are phenomena such as drought and heatwaves that impact communities and ecosystems across Australia. The immense suffering of Australian farmers during drought periods has long been a point of socio-political tension across the nation. Although farmers' livelihoods often take centre stage in discussions of drought, in the background there is always the nagging matter of the unsustainability of Australian settler-colonial agriculture and the immense devastation it has caused to native species and habitats since European colonisation. Whilst drought is far from an uncommon trope of Australian settler-colonial culture, the success of *The Dry* suggests that

Harper's intense focus on it has struck a nerve with readers, both in Australia and overseas. Notably, early in Harper's novel it is established that the drought conditions in rural Victoria described in the narrative are not of an ordinary kind: they are "officially the worst conditions in a century" due to "El Niño", the Melbourne weathermen say (Harper 2017a, 1). Statements such as these, made by meteorologists in their "air-conditioned studios" (1), stand in stark contrast to the harsh reality experienced by farmers out on the land. Whilst Harper does not directly link this drought with climate change, it would be difficult for some readers to ignore the relevance of the climate crisis in this context, especially given that recent intense droughts in Australia have been strongly linked to human-induced climate change (Jones et al. 2013, 2). The positioning of Falk as a witness to the crimes of Kiewarra and the associated environmental crisis is significant in this regard.

From the beginning of the narrative, Falk is portrayed as a spectator to the drought and the crimes committed during it—he is an outsider who "had chosen his life in Melbourne" (Harper 2017a, 151)—but, importantly, he is also forced to choose this city life due to his alleged involvement in the death of Ellie Deacon when he was a teenager, and he is also friends with the alleged murderer, Luke. Falk is therefore a witness, a potential criminal and, as an investigator, a seeker of justice, representing Rigby's conceptualisation of the ecoprophetic witness who "is both implicated in and wounded by … wrongdoing" (2009, 178). While Falk begins the novel as an outsider from Melbourne, he quickly develops a newfound emotional connection to his childhood town and the suffering linked with the murders and the drought. Visiting the river that he and Luke swam in as children, Falk is shocked to discover that it is dry; he "clamber[s] into the cavity, hands and knees scraping against the baked bank" (Harper 2017a, 106). This moment is a dark epiphany, highlighting the extent of the drought and how much his beloved childhood places have been so drastically marred by both murder and ecological catastrophe. This realisation more broadly represents that of the settler-coloniser coming to terms with their role in the climate crisis, and the implications this has for current colonial systems of domination—such as still-widespread Anglo-agricultural practices—that continue to enact environmental devastation and nonhuman violence; coming to terms with this devastation means accepting that current practices need to change if any real action on climate change is going to be possible.

Harper also draws attention to the role that the Australian Federal Government is, or rather is not playing in drought relief. Falk's first interaction with a Kiewarra local on his return to town for Luke's funeral involves another funeral attendee berating the government for "letting things get this bad" (Harper 2017a, 6). In his critique of the government, this man disregards Luke's part in the murders that, at this point in the narrative, the town believes he has committed. He states, "you can hardly blame the poor bastard" (Harper 2017a, 6); to this local, criminal culpability belongs with the Australian government, not the local farmer supposedly driven to desperation. Significantly, such a belief positions the Australian bush and associated drought as a central cause of rural violence. This relates to Gelder and Weaver's comments about the perceived role of the Australian natural world in aggravating crime (2008, 6), enabling people—often men—already under pressure in an unforgiving natural setting to commit atrocious crimes without shouldering the culpability.[4] This local's view suggests that it is no wonder regular men like Luke Hadler commit murder because drought conditions effectively make them do it, representing what Knight describes as "a curious displacement of human agency onto the land" (2006, 20).

Such a displacement relates to Gibson's concept of Australia's postcolonial badlands, summarised by Potter as "spatial containers that encyst a culture's refuse and confronting madness, and—because of this quarantining—condemn a particular place to repetitions of 'bad'" (2019, 104). In Australia, these areas are those that, as Potter describes, "become popularly synonymous with violence, unpredictability, excess and fear" (2019, 104). Their categorisation as "bad" also calls upon elements of noir, condemning such places to exist only as the settings for settler-colonial crime. In both *The Dry* and *Scrublands*, these "bad" places may be abandoned or isolated farms, agricultural townships on the verge of economic and/or environmental collapse, or unpredictable native Australian habitats experiencing drought and bushfire one moment, and flooding the next. As aforementioned, these Australian badlands and their history of settler violence can be likened to an "immense, historical crime-scene" (Gibson 2002, 1).

[4] This denial of culpability is a common thread in the reporting of murder-suicides in Australia. In particular, media reports about Peter Miles and Geoff Hunt, two men separately accused of murder-suicide involving their own families, described the perpetrators as good men and subsequently attracted widespread criticism from readers and journalists. See Media Watch 2018; Ford 2018.

Just as these badlands and the violent events that take place within them reflect Australia's bloody past in their "challenging environments and isolated townships" (Potter 2019, 105), so too does the setting of Harper's novel reflect the struggle of settler-colonisers to live more harmoniously alongside the Australian natural environment. As indicated above, there is a perception that the harsh drought that engulfs the Kiewarra township exacerbates violent tendencies and paranoia. This is made especially evident when Falk and police sergeant Greg Raco absorb the starkness of the place where Luke Hadler is found dead:

> There was a pocket of eerie silence as invisible birds were momentarily stilled by the sound of his voice.... It was fractionally cooler here, shaded on all sides by a sentry line of ghost gums. The road was completely hidden by thick growth. Something in the bush rustled and scurried away. The yellow earth was baked solid. (Harper 2017a, 83)

Interestingly, this passage highlights the presence of nonhuman life, even in drought—the birds, the trees and the undergrowth—and yet, this place surrounded by bush is still portrayed in a negative light: Aaron describes it as a "pretty miserable place to spend your last moments" (Harper 2017a, 83). Gelder and Weaver's concept of the Australian bush as responsible for instigating violent crimes is relevant here (2008, 6), the suggestion being that Luke Hadler died because the overbearing power of the bush willed it so, raising questions about the capacity of settler-colonisers to live peacefully and productively on land that still feels so unfamiliar and unforgiving to them.

Harper's narrative suggests that the nature of crime is always shifting in these landscapes that will continue to experience a combination of extreme weather events, as the impacts of climate change worsen over the years. Intense heat and prolonged drought result in the redundancy of typical human weapons, with power instead being assigned to other, more unexpected objects. When school principal Scott Whitlam is discovered as the murderer of the Hadlers, he escapes into the nearby bush and is pursued by Falk and Raco. He is found holding something that is "a thousand times worse than a gun"—a lighter (Harper 2017a, 315). This tense moment is both ridiculous and horrifying; the drought has granted a power to Whitlam's lighter that in less extreme climatic circumstances it would not possess. Gelder and Weaver refer to the anxieties depicted in early Australian crime fiction as emanating from "a sense of the colonial landscape as a fluid, transient place" (2008, 4), as well as settler-colonial fear regarding flammability and arson in the Australian bush. One year, a lighter in the bush may be relatively harmless, but in another, drought-ravaged year, it becomes deadly. Its use as a murderous tool in *The Dry* suggests that environmental catastrophe and intra-human violence have converged on one another. This emphasises the role of the climate crisis and the resulting intense droughts and bushfires in exacerbating social and political tensions in Australia, particularly in rural and still largely white, settler-colonial communities.

In the depiction of such tensions, water—seemingly the opposite of bushfire and drought—also plays a key role in Harper's novel. Specifically, the novel prominently features a river at the centre of one of the crimes committed. Ellie Deacon, Falk's high school friend, was found dead in the Kiewarra River when she was sixteen, resulting in Falk and his father being forced out of town by those who believed Falk responsible for the murder. As described earlier, on returning to the river while investigating the Hadler murders, Falk finds to his horror that it is bone dry. Although unsurprising given the prolonged drought, the reality of it hits Falk in an unexpected way:

> Struggling to accept what his eyes were telling him, Falk clambered into the cavity, hands and knees scraping against the baked bank. He stopped in the dead centre of the river, in the open void where the heavy ribbon of water had once been deep enough to close over his head. (Harper 2017a, 106)

It is this same water that gave life to Falk and Luke's childhood, and took away the life of Ellie Deacon, "greedily invading her body until there was no room left for the girl herself" (Harper 2017a, 106). The transient nature of the river both physically and symbolically speaks to Gelder and Weaver's comments on early Australian crime fiction; how the genre "often turns on anxieties that flow from a sense of the colonial landscape as a fluid, transient place" (2008, 4). The Kiewarra River is such a place, physically flowing with water one year but empty the next, and symbolically capable of facilitating life, but also taking it away. Harper's description of the water as agential in this context is significant; Ellie was murdered by human hands, but it is the water that "forced its way down [her] throat" (Harper 2017a, 106). The nonhuman is ultimately responsible here, regardless of human involvement. This provides an example of the way in which Harper somewhat discredits the quintessential settler-colonial narrative of nature as something to control (the river is depicted as having an agency of its own). Yet, this passage also upholds the idea that Australian nature is somehow vindictive or, at the very least, a facilitator of violent ends to human lives.

As with other works of Australian ecofiction discussed in this book, *The Dry* is very much on the borderline of both challenging and conforming to typical settler-colonial understandings of nature and the nonhuman. Harper's narrative still demonstrates a problematic ambivalence towards the natural world—like the depiction of drought as exacerbating the threat of domestic violence, the river's agency does not facilitate a hopeful relationship between the human and nonhuman in the same way that the agency of, for example, the fig tree does in McKinnon's *Storyland*—and yet it is still a work that strongly engages with themes of ecological crisis and settler-colonial responsibility. Like *Scrublands*, *The Dry* is a novel that, while not an overt commentary on the climate crisis, encourages a dialogue between ideas of criminal responsibility, intensifying climate conditions, and drought and bushfire in Australia.

The Dry makes an important link between settler-colonial approaches to drought and the treatment of individual nonhuman animals, particularly livestock. This is poignantly represented in the prologue when the narrator states that "there was little difference between a carcass and a corpse" (Harper 2017a, 1). This passage speaks of how "the farmers of Kiewarra levelled their rifles at skinny livestock. No rain meant no feed. And no feed made for difficult decisions" (Harper 2017a, 1). As the prologue concludes, the meaning of the introductory sentence becomes clear: based on the description of "blood pooled black over tiles and carpet" and "a child's scooter … abandoned" (Harper 2017a, 2), the reader learns that these carcasses are, in fact, human corpses. Such a comparison suggests a blurring of boundaries between what violence may be considered ethical in a settler-colonial landscape—for example, the killing of farm animals—and what is unethical—for example, the killing of a human. Prolonged drought is depicted as being responsible for this moral clouding, resulting in desperate men committing horrific crimes when they can no longer make ends meet, as the man at Luke Hadler's funeral suggests.

Such symbolism implies that the human (in particular, the settler-coloniser) can be both implicated in and, problematically, freed from criminal responsibility in the context of environmental crises. As Rose describes it, "we are spectators in the unmaking of the world" (2013, 211), but we are also, as Morton puts it, "narrators of our own destiny" (2016, 9), suggesting that settler-colonisers are both observers and perpetrators of violence committed against the natural world. Notably, there is no mention of the party responsible for the murders in this opening passage; the only other living animals present alongside a crying baby (Luke's youngest child, left alive) are the "blowflies" swarming the dead bodies (Harper 2017a, 2). This absence of immediate human culpability, although typical of a "whodunnit" crime narrative, is indicative of a similar denial of settler-colonial culpability for the crime of anthropogenic climate change that has resulted in such intense drought. Instead, the blowflies represent this culpability, suggesting that it is the state of the natural world in drought—often typified by swarming flies—that has facilitated these murders.

Throughout the narrative, Harper likens the townspeople of Kiewarra to various farm animals, describing pub goers "star[ing] with bovine blankness at greyhound racing on TV" (2017a, 49–50) and Grant Dow, cousin of the murdered Ellie Deacon, as "piggy" (59). These discriminatory remarks suggest a distancing between certain townsfolk and the horrifying impact of drought on nonhumans, particularly those associated with settler-colonial agriculture. These animals suffer during drought, but they are often not granted the same respect as those humans, such as Luke Hadler, who suffer in the same way. This highlights the complex relationship Australian settler-colonisers have developed with agricultural animals over the years, treating these nonhumans adequately when humans rely on them for their livelihoods and then treating them with violence when drought and bushfire take their toll on the landscape. This is reflected in the aforementioned passage in which farmers are described as "level[ing] their rifles at skinny livestock" (Harper 2017a, 1), forced to euthanise those animals out of both necessity and mercy, as well as in Falk's observation of "animals [that] lay dead in paddocks" (Harper 2017a, 106). Descriptions published in news stories across the country are shockingly similar to Harper's fictional portrayal, particularly during Australia's 2019–2020 Black Summer. Journalists Sophie Meixner and Tom Lowrey write of how, in an area of New South Wales particularly hard-hit by the bushfires, "animal carcasses lin[ed] the sides of the road, with farmers beginning the slow, difficult and grim work of loading the bodies onto the trays of utes"; many surviving animals also had to be euthanised (2020). These mass livestock deaths indicate one horrifying way in which the climate emergency will continue to impact Anglo-agricultural livelihoods into the future, and crime novels such as *The Dry* are potentially one means by which readers can engage with ethical questions surrounding the plight of nonhumans in these situations.[5]

It should be noted that it is both these animals, and the affected native nonhumans, whose backs the Australian agricultural industry have been built on. In *The Dry*, drought has resulted in large numbers of livestock deaths. When visiting the local school, Falk notices drawings by young children showing the devastation of the drought:

> There were stick figure families in which every face had a crayon mouth turned downwards. A painting of a cow with angel wings. *Toffee My Cow in Heaven*, the shaky caption read. In every attempt at landscape, the paddocks were coloured brown. (Harper 2017a, 87)

The sadness of Kiewarra's human families is not only reflected in their stick figure faces, but in the cow gone to heaven and the brown landscapes devoid of life. The wilting wattle sapling that is planted at the school in honour of the murdered Hadlers appears "sickly", unlikely to "stand a chance" in the intense heat of the drought (Harper 2017a, 85). This, too, connects human death to the death of a nonhuman, signifying that with the loss of one comes the other. It also expands upon the enduring association between Australia's natural colour palette (often browns and greys) and Anglo-agricultural failure—the knowledge that if the fields are not green, then grazing livestock cannot survive.

Furthermore, there are strong similarities drawn between the mistreatment of nonhumans and intra-human violence in Harper's novel. The non-native European rabbit plays a significant role in this respect, which is brought to light when Falk reflects on an encounter between Luke, himself and a baby rabbit that they find in overgrown grass as teenagers. In the present narrative, an adult Falk has just hit a rabbit with his car, feeling "the sickening thud under the wheels" (Harper 2017a, 31) before he has time to avoid it. Comparatively, it is suggested in Falk's flashback that the teenage Luke may have intentionally killed their rescued rabbit:

> He had one hand in the box. Luke's head snapped up as Aaron approached and he snatched his hand out…. "It died," Luke said. His mouth was a tight line. He didn't meet Aaron's gaze. (Harper 2017a, 31)

[5] It is worth noting, though, that there is primarily an emphasis on non-native animals, such as farm animals, in this novel, with little consideration granted to native wildlife affected by drought and bushfire.

A connection is suggested here between Luke's presumed past violence towards an animal, and Falk's suspicion that Luke may be responsible for both Ellie Deacon's murder, and that of his own family. The sudden killing of a baby rabbit seems completely at odds with Luke's love for his childhood dog (Harper 2017a, 154), suggesting that perhaps Luke was not responsible for the rabbit's death, nor that of his family's. Comparatively, Mal Deacon—Ellie's father, who *is* responsible for her murder—exhibits extremely violent tendencies towards nonhuman animals. Mal is described as exercising "a brutal hand" on his flock of sheep come shearing time, leaving them "a staggering, bleeding mess" as he shears them "too close to the skin" (Harper 2017a, 67). Falk also recalls how Mal hits Luke's dog with his car, leaving a fourteen-year-old Luke crying over his much-loved pet "as he'd cradled it by the roadside" (Harper 2017a, 154). These encounters between human and nonhuman echo a similar sentiment to Rose's concept of Anthropocene noir. Many characters in *The Dry* are spectators of "the misery of numerous and diverse forms of life" (2013, 211), extending the sorrow associated with murdered humans and their families to the desolation of nonhumans implicated in human violence and environmental crises such as drought.

However, the representational aspect of these nonhuman portrayals is perhaps problematic. The relationship between the nonhuman and intra-human violence, as depicted in *The Dry*, can be interpreted in two ways: symbolically and one of cause and effect. I argue that Harper in fact represents both in her work. Firstly, she presents the above suffering of animals as symbolic. This is particularly significant in the case of Luke Hadler and Mal Deacon, as the above instances of animal cruelty could be viewed as proleptical for the human murders that they are later accused of committing. Of course, violence committed against the nonhuman is often understood as a sign that a person is capable of intra-human violence as well.[6] While this psychological link is important to recognise, it is also representative of how animal cruelty is perceived as a sign of worse behaviour, when the act of animal cruelty alone is horrific as it is. This positioning of animal cruelty as a crime that comes before intra-human violence is nothing new, but does deflect from the experience of the nonhuman in these circumstances and also depicts animals as "scarcely animals at all but schematic elements in an aesthetic or psychological design" (Scholtmeijer

[6] For a recent study on the link between animal cruelty in childhood and violent crimes committed by adults, see Hensley et al. (2017).

1993, 259, cited in Donovan 2016, 95). Philip Armstrong similarly describes this as "reduc[ing] the animal to a blank screen for the projection of human meaning" (2008, 3, cited in Donovan 2016, 95). While I would argue that Harper largely relies on the symbolic power of the nonhuman in her novel, she also utilises connections between cruelty to animals and intra-human violence in a way that highlights how these events inform one another in reality. Events such as bushfires lead to increased animal deaths, as detailed above, and while many of these deaths may be deemed humane euthanasia, the cruel conditions in which these animals are kept in the lead-up to their deaths (lack of feed and water, and exposure to extreme heat) would be partly avoidable if human-induced climate change was not a reality. With livestock deaths come farmers' desperation to make ends meet, and so the cyclical nature of violence experienced by nonhumans and humans is perpetuated.

6.3 The Justice of Bushfire in Chris Hammer's *Scrublands*

Hammer's debut novel *Scrublands* is set in a fictional Australian country town in the Riverina region of New South Wales. Described by Laura Wilson as a work of "outback noir" (2018), *Scrublands* emphasises the oppressive Australian heat and seemingly endless drought. Significantly, the protagonist of *Scrublands* is not associated with law enforcement but is rather a journalist chasing a story. Former war reporter Martin Scarsden travels to Riversend on the first anniversary of a tragic mass shooting that occurred at the town's church, committed by the local priest. After speaking with locals, Martin realises that the existing narrative regarding the shooting requires further investigation. The bodies of two murdered backpackers are then discovered on a farm outside of town, raising more questions about which townsfolk have secrets to hide. Hammer has also released two sequels to *Scrublands* titled *Silver* (2019), which sees protagonist Martin Scarsden return to his seaside hometown, Port Silver (a similar plot point to Falk's return to Kiewarra in *The Dry*) and *Trust* (2020), in which Hammer shifts to the urban focus of Sydney. As a journalist, Hammer has also published two non-fiction works prior to establishing himself as a fiction author. One of these texts, *The River* (2010), notably investigates the devastation of the Murray-Darling Basin and the communities and ecosystems reliant on it: an issue that sits at the centre of *Scrublands*.

Like *The Dry*, there is a focus on broader nonhuman forces in this text through depictions of phenomena such as drought and bushfire. Known for its combination of forested wetland and dry land habitats (Zhang et al. 2017, 331), the Riverina area of New South Wales has experienced periods of intense drought with severe flooding often occurring in between (Smee 2019). In *Scrublands*, the heat of the drought is perceived as a nonhuman entity with an agency of its own; it is "insistent" and Martin "can no more ignore it than he can ignore life itself" (Hammer 2018, 462). As Martin and Robbie Haus-Jones, a local police officer, drive through a fire that is threatening properties in the titular scrublands outside of Riversend, Martin notes that "the world around them grows increasingly apocalyptic, the sky closing in, the light fading, ash falling, some of it glowing orange at the edges". It is notable that the bushfire is compared to a non-natural object: a "freight train [that] smashes into the back of the house" (Hammer 2018, 101). There is a suggestion here that the fire is something more than a natural occurrence; it represents what Knight describes as "a curious displacement of human agency onto the land", a force that is enacting justice for lands stolen and devastated by the settler-coloniser (2006, 20).

The bushfire is also depicted as being partially responsible for revealing the remains of two missing backpackers on a farm outside of town. This alludes to what Knight describes as a kind of zero-policing crime story ("where the just outcome is achieved without a detective and where the moral resolution is often in some way natural, as when execution occurs through fire or flood" (2006, 24)). As cited earlier, Knight describes how bushfires are often portrayed as "agents of justice" in these zero-policing crime stories (2006, 20). While the bushfire in *Scrublands* does not bring about justice in a direct sense, it does reveal the bodies of the missing backpackers and the murders are later linked to Jamie Landers and Allen Newkirk, two teenage boys whose fathers were murdered in the church shooting. In this sense, justice for these murders is indirectly achieved through fire, as it is the means by which the bodies are first revealed. As in Harper's work, a link is drawn between settler-colonial discourse and an agential kind of nonhuman phenomena that is to blame for exacerbating human violence through drought, and yet is also responsible for dispensing justice.

Related to this is the fact that the title of Hammer's novel refers to the Scrublands—an area outside of town described by Robbie as "shit country. Mulga scrub. Hundreds of square kilometres of it" (Hammer 2018, 75). This naming of the text after a natural place (the Scrublands) rather than a human-made place (Riversend) is significant. This "shit country" (Hammer 2018, 75) is arguably surviving better than the town and people of Riversend, the mulga scrub resisting the intense heat, although the imminent threat of bushfire is palpable. This emphasis on the natural rather than human place in the novel's title implies that it is this area that is most associated with the crimes of the novel, despite the fact that many other crimes take place in the centre of town, including at the church. This is a further example of the ways in which Australian crime fiction is still conforming to this idea of the bush as responsible for "the kinds of degradation and debasement that enable criminal impulses to flourish" (Gelder and Weaver 2008, 6). Instead of emphasising the human crimes of the built environment, Hammer more strongly highlights the violence and destruction that takes place in the bush.

Significantly, though, such events are subtly linked to climate change in Hammer's novel, suggesting that while upholding some traditional aspects of settler-colonial discourse regarding the Australian natural world (the reduction of the nonhuman to "a blank screen for the projection of human meaning" (Armstrong 2008, 3)), *Scrublands* also demonstrates an awareness that nature as we know it is not behaving as it should. Martin likens Riversend's claustrophobic climate to "a Pacific atoll with rising sea levels gnawing at its shores" (Hammer 2018, 251), and the following passage alludes to the relationship between drought and broader environmental crises occurring elsewhere in the world:

> He knows that somewhere in the world there must be clouds.... Somewhere it is raining; somewhere it is pelting down. There will be floods and landslides and hurricanes and monsoons.... More water than you can imagine, more water than you could ever want. Somewhere, but not here. Here there are no clouds and no rain. The drought can't last forever; he knows it, everyone knows it. It's just become hard to believe. (Hammer 2018, 443)

Situations such as these have, of course, occurred very recently in reality. As Australians bear witness to climate-related disasters overseas, many simultaneously experience similar disasters at home—disasters that, like the drought in *Scrublands*, are almost "hard to believe." For example, while extreme flooding and landslides killed dozens of people in Jakarta in early 2020 (Renaldi and Handley 2020), Australians were experiencing the worst bushfire season on record. As Rose describes, it may be that Australians are fast becoming "spectators of our own demise" (2013, 211), watching as these unbelievable environmental catastrophes become reality and yet failing to implement effective action quickly enough.

Knight argues that a particular trend emerged in Australian crime fiction in the latter half of the nineteenth century: the abandoning of police as crime-solvers (2006, 24). The protagonist of *Scrublands* is not a police figure, but rather a journalist following a story, suggesting that Hammer's text subscribes to the anti-authority sentiment prevalent in settler-colonial culture. Whilst Knight claims that the detective protagonist was reintroduced to Australian crime in the 1980s (2006, 24), the trend of portraying civilian crime-solvers has re-emerged in the twenty-first century. As well as Martin Scarsden of *Scrublands*, other popular Australian crime novels of the past several years include non-police protagonists: Brandi's *Wimmera* (2017) and *The Rip* (2019), Viskic's *Resurrection Bay* (2015) and Harper's *The Lost Man* (2018) are some examples. What this may indicate about the relationship between the human and nonhuman in these novels is interesting and relates to the anti-authority "everyman" prevalent in Australian colonial literature.

The strong emphasis on an "everyman" in settler-colonial cultural expression has led to an admiration for "everyday heroes". Jeanine Leane describes the "handsome, care free, anti-intellectual bushman, apathetic and unaware of the wider world around him" as an identity that was manufactured by settler-colonisers following Australian Federation (2010, 32). The writing of non-authority protagonists into Australian crime fiction speaks to this, suggesting that whilst texts like Hammer's may provide opportunities for readers to understand the impacts of environmental catastrophe, these narratives still very much sit within the Australian settler-colonial literary tradition and draw on anti-authority sentiments prevalent within settler-colonial discourse. This is problematic when one considers the environmental destruction caused by colonisation in Australia; the text potentially acts to deny this fact by upholding certain settler-colonial stereotypes that encourage a view of the Australian natural world as unforgiving and in need of taming by the coloniser—for example, Martin's perception that the bushfire "wants to come in and eat us" (Hammer 2018, 101) and that "the sun hangs over Riversend like a sentencing judge" (Hammer 2018, 457). Similarly, the "self-contained" layout of Riversend is what seems to bring stability to the town that is otherwise "adrift on the vast inland plain" (Hammer 2018, 26). The centrality of the settler-colonial everyman in *Scrublands* also possibly explains the lack of Indigenous characters depicted in the text, as well as an absence of meaningfully depicted female characters. As Leane contends, the exclusion of "women, urban people, immigrants and Indigenous people" from many settler-colonial narratives is a result of emphasising the settler-colonial bushman identity (2010, 32).

This latter point regarding the characterisation of women in these narratives is particularly pertinent in an analysis of *Scrublands*. Mandalay, or Mandy, Blonde is a problematic depiction of the love interest archetype often seen in crime, portrayed by Hammer as a young and vulnerable single mother who is exploited by protagonist Martin (throughout the novel, he consistently presses Mandy for personal information for the purposes of his story while also pursuing a romantic relationship with her), Byron Swift (the father of her child, who never reveals his true identity to her and romances other women behind her back) and Harley Snouch (accused of raping Mandy's mother). When Martin first meets Mandy, he is "disconcerted … by her beauty" and notes her "husky voice" (Hammer 2018, 10), immediately positioning her as a potential future conquest. The text's endorsement of this patriarchal narrative is problematic as it is, but especially problematic in the context of settler-colonialism, drought and environmental crisis. There is an implicit suggestion here that Australia should continue to rely on white, male leadership in times of intense drought and bushfire; a kind of leadership that has recently proven so unreliable.[7]

It is worth noting here that some Indigenous Australian fiction utilises certain tropes of mystery and noir fiction in a way that often opposes problematic settler-colonial perspectives of nature and the nonhuman. Melissa Lucashenko's *Too Much Lip* (2018) and Nardi Simpson's *Song of the Crocodile* (2020) are two recent examples of Indigenous Australian literature that focus on unsolved mysteries stemming from the violent settler-colonial past of both novels' settings. Julie Janson's *Madukka The River Serpent* (2022) was explicitly marketed as First Nations crime fiction and, as with Lucashenko's and Simpson's works, is explicitly concerned with the links between past colonial crimes, environmental exploitation, and

[7] The leadership of Prime Minister Scott Morrison during the 2019/2020 bushfire disaster has been strongly critiqued, as summarised by James Walter in *The Conversation* (2020).

present-day crimes and prejudice in the Australian justice system (see Fetherston 2023). Similarly, Philip McLaren's crime fiction—in particular, *Scream Black Murder* (2002)—features Indigenous detectives investigating crimes committed against both Indigenous and non-Indigenous Australians, highlighting the differences in media attention and police interest assigned to such cases. The works of these First Nations authors demonstrate that crime tropes have been and continue to be used to undermine the settler-colonial project in Australian literature, highlighting the possibility of utilising such tropes without conforming to settler-colonial stereotypes as Harper's and Hammer's novels do.

Another possible interpretation of the everyman trope used in modern Australian crime fiction is that the everyday Australian is likely to be implicated in environmental catastrophe; during bushfire events throughout Australia's history, volunteer firefighters and civilians have commonly been on the frontline of these disasters. Whilst law enforcement plays a role in the national discourse surrounding these catastrophes, there is a larger focus on the "everyday heroes" of Victoria's Country Fire Authority and the Rural Fire Service of New South Wales, for example, and the everyday people—often farmers and rural townspeople—whose homes and livelihoods have been destroyed. In a review of the media coverage of various Australian bushfire disasters, Susan Yell notes the prevalence of specific iconic images that appeared in this coverage, such as firefighters battling fires and survivors visiting the ruins of their properties (2010, 114–15), suggesting that volunteers and an Australian "everyperson" play key roles in cultural perceptions of bushfire. In *Scrublands*, it is much the same, with reporter Martin helping to fight an out-of-control blaze in order to save a local man's property (Hammer 2018, 100–105).

Hammer also highlights how the quintessential rural Australian town, made up of "everyperson" citizens, can resort to crime in the dire circumstances of extreme drought. Martin discovers that Byron Swift, the priest who commits the mass shooting at the local church, had been funnelling drug money to those in Riversend who desperately needed financial assistance in the wake of severe drought. Love interest Mandy Blonde, who is also a local business owner, tells Martin:

> ... we all knew there was money. For the footy team, for the youth group, for families that were doing it tough. To help with the fire brigade, with the services club. We were in it together—the drought was getting worse, closing in—we didn't ask questions. (Hammer 2018, 457)

Byron has also been living under a fake name to escape his past as a war criminal in Afghanistan, for which he is attempting to make amends. The stagnant nature of the drought-prone landscape stands in contrast to the fluidity of Byron's identities. Both he and others in Riversend, such as the character Harley Snouch whose crimes are both confirmed and denied throughout the narrative,[8] maintain these fluid identities in order to commit and then absolve themselves of crimes, lending credence to Gelder and Weaver's idea of the constantly unfolding colonial identities that were especially present in early Australian crime fiction; as they state, "characters continually slip out of view, changing their identities at will and sometimes even disappearing altogether" (2008, 4). Crime, in this case, is helping Riversend combat the worst impacts of the drought crisis, and Byron's identities are manifold: he is portrayed as an outsider (a non-local with a fake name) and a criminal (a murderer), but also as a bringer of justice, when the reader finally learns of his motivations for the shooting (wanting to protect the women of the town from the violent Craig Landers (Hammer 2018, 465)). Like Aaron Falk in *The Dry* who acts as an "ecoprophetic witness" (Rigby 2009, 178), Byron's shifting identities show him to be a spectator witnessing the struggles of those living with the impacts of climate change and also as someone who is implicated in the crimes from which he is trying to save people.

This relates to the fluid nature of ground, which, as discussed by Potter, is always "in movement" (2019, 103). The criminal activities that occur in Riversend are both ruinous and necessary due to the drought's capacity to destabilise these grounds. When neatly summarised in the novel's penultimate chapter, the broad array of crimes committed in the context of desperation and drought speaks to an overarching sense of unsettlement in Riversend:

> Four different crimes, all taking place in and around the same drought-ravaged town, all separate but all interlinked, driven by greed and hate, guilt and hope: the drug operation … the murder of the Germans … the shooting at St James … and Harley Snouch, attempting to expunge rape with fraud. Four crimes, all seeded by violence from the recent or distant past. (Hammer 2018, 472)

[8] At various points in the novel, different characters either endorse or dispute that Snouch raped Mandy's mother.

These interlinked crimes reveal a similar connectedness between drought, anthropogenic climate change and settler-colonial environmental destruction. All can take place separately, but all are also concerned with one another, each exacerbating the other and often indirectly leading to the intensification of intra-human violence.

This indicates a feeling of uncertainty that is particular to postcolonial place, and subsequently relates to Gibson's "postcolonial badlands" (2002) and Morton's discussion of noir and ecological crisis which contends that humans are both observers and perpetrators (2016, 9). Hammer's novel positions the land on which these crimes occur as ceaselessly oscillating between steady reliability and unstable ground that exists "in the midst of things unfolding" (2019, 104), suggesting that ecological catastrophe is destabilising the position of the settler-coloniser and bringing their culpability to the fore.

As in *The Dry*, water as a recurring motif plays a key role in destabilising the supposedly solid ground of settler belonging in Riversend. At one point, Martin reflects on Heraclitus' aphorism: "that a man cannot step into the same river twice" (Hammer 2018, 460). "Does it hold for waterless rivers?" (Hammer 2018, 460), he ponders, concluding that the answer is no. Despite his foray back into a war zone—this time, not that of Gaza but rather the war zone of drought, deception and murder—Martin is no longer the same person; he is "being reborn" (Hammer 2018, 461), even in the absence of symbolically cleansing waters.

The conclusion to the novel emphasises the literal and symbolic value of water, as the much-awaited rains finally reach Riversend, cleansing the town of the lingering remnants of its crimes:

> The plain runs on forever, the sun omnipotent, the air dry, but today is different. For marching across the horizon, as if painted onto the blue backdrop of the sky, is a front of clouds, dark and purposeful, a rare low-pressure system penetrating Australia's interior for once… (Hammer 2018, 476)

When the rain finally falls over Riversend, the rolling thunder is likened to "a church bell" (Hammer 2018, 481). Now the site of so much violence, St James Church may no longer be the sanctuary it once was. However, a natural event—in this case, rain—provides redemption in a way that human institutions cannot: the rain offers a shift towards "[l]ife. At long last" (Hammer 2018, 481), rather than death. This metaphor alludes to the Romantic traditions inherent in settler-colonial understandings of Australian nature—that is, the ideal of nature as "spiritualised" (Oerlemans 2004, 33) providing symbolic opportunities for settler rebirth and redemption rather than existing as a physical entity with stories of its own.

Following on from this idea of water and rebirth, it is therefore notable that another distinctive theme of the novel is religion. Many of Riversend's townsfolk find it difficult to believe that Byron Swift could have been responsible for a mass shooting given his pious nature. Codger Harris, a local hunter and one of few people who live in the Scrublands, tells Martin that even when Byron went hunting, "he used to say a little prayer before he started shooting and a little prayer afterwards, for the animals he killed" (Hammer 2018, 84).(Here, though, we return to the idea of fluid identities—as Martin observes about Byron, "Here he is, preaching love for all living things, tolerance and forgiveness, and then he's out killing things, shooting animals in the Scrublands" (Hammer 2018, 220). These seemingly disparate ways of being reinforce Gelder and Weaver's comments regarding the ever-shifting state of settler-colonial identities as depicted in early Australian crime fiction (2008, 4), but also reveal something about Byron's role as priest: in order to be a good leader within the community, he must be sinful, obtaining funds through the drug trade but redistributing them to enact positive change in Riversend in the face of environmental catastrophe. Byron also does this in an effort to redeem himself for past sins committed in Afghanistan—effectively committing one crime in order to alleviate the guilt of another.

Richard Bohannon contends that religions can contribute to conversations about environmental issues in significant ways via their narratives of meaning-making, allowing for understandings of how humans and non-humans relate to one another and "how we should act within the world" (2014, 3). As demonstrated by Byron's genuine desire to assist the people of Riversend, Christianity has a role to play in diminishing the impact of environmental injustice, both on the basis of combatting social injustice, which many religious groups already play a role in, and by recapturing "the sense of sacredness and intrinsic worth" of the creation of nature (Kearns 2014, 299).

However, it should be noted that Christianity still largely represents the settler-colonial narrative of human and nature as separate. There have been recent interpretations of Christian scripture (particularly of Genesis) indicating strong support for human stewardship of nature and the non-human, but Sylvia Mayer contends that the long-held reading of the Bible that maintains human–nature dualism "has only started to lose its conceptual force and is still used as a legitimization of economic practices that operate worldwide" (2006, 115). As summarised by Rigby, historian Lyn White (1996) contends that it is vital to understand the ways in which "the dominant religious traditions of the West" are responsible for "legitimising that ruthless technological manipulation and economic exploitation of the earth, which has today reached a level that would have been unimaginable, and quite possibly appalling, to the authors of Genesis" (2008, 4). In this sense, Hammer is also upholding the settler-colonial sensibilities regarding Australian nature as a kind of aesthetically pleasing, agriculturally dominated creation (such as the sweeping paddocks of Australia's farming landscapes) rather than the reality of it as vastly different to the European pastoral aesthetic. Settler-colonial domination of nature and the nonhuman is also affirmed by Hammer in his depiction of the sadistic treatment of a missing cat found dead by Martin in the abandoned pub (Hammer 2018, 232) and the disregard for the wallabies, run over by Martin and Robbie as they drive to save an alleged criminal from the bushfire (Hammer 2018, 99).

When Martin encounters St James Church for the first time, he notes how the sun "sends shafts of golden light through the branches of the river red gums" (Hammer 2018, 32). This light suggests a sacredness to this place and, whilst not altogether unexpected given the spiritual location, is perhaps shocking given that this was the setting for a violent mass murder. This again reveals the uncertainty and fluidity of postcolonial place and ground, always oscillating and constantly "in movement" (Potter 2019, 103). The events that occur on this ground are both sacred and violent, and the nature of drought seems to destabilise these grounds even more so.(

6.4 Conclusion

Both *The Dry* and *Scrublands* demonstrate the potential of Australian crime fiction to communicate the threats to both human and nonhuman life should the impacts of the climate crisis continue to intensify. While neither text goes into explicit detail about the links between intense drought and bushfire seasons in Australia and anthropogenic climate change, the publication of *The Dry* and *Scrublands* in 2016 and 2018, respectively, is notable; their release was situated between the Black Saturday bushfires of 2009 and the devastating Black Summer bushfires of

2019 and 2020. In 2017, Australia also saw the beginning of an intense drought season which has been linked to climate change (CSIRO and Bureau of Meteorology 2020). In this context, Harper's and Hammer's portrayals of the impacts of drought and bushfire are important, introducing readers to the struggles of Australians (although primarily white, settler-colonial Australians) and nonhumans living under these impacts through the use of familiar crime tropes that may engage a broader audience than literary Australian climate fiction.

The representation of the nonhuman in these texts alludes to the ways in which settler-colonisers are largely responsible for the vast changes to Australia's natural environments and the suffering experienced by nonhumans as a result, suggesting an interconnectivity between broader ecological crises and the unsustainability of the settler-colonial project. Both *The Dry* and *Scrublands* engage with what Potter considers "the colonial obsession with stabilized ground" (2019, 103). Ever-shifting identities, unreliable human and nonhuman forces, and the irrevocably devastating impacts of ecological crises are major focuses of these narratives, and the authors' emphasis on this latter theme suggests a critique of current settler-colonial systems (in particular the sustained dominance of Anglo-agricultural practices). My discussion of Rigby's ecoprophetic witness as represented in these texts also contends something similar, putting forward the idea that settler-colonisers are witnessing and affected by a catastrophe that they have had a hand in bringing about.

However, whilst Hammer's and Harper's depictions of the nonhuman partly represent a criticism of Australian settler-colonialism, their works also uphold a quintessential view of the Australian bush as the reason for exacerbated levels of violence in rural communities, subsequently denying human culpability for both domestic crimes and broader ecological crimes. Merely setting these violent stories in rural Australian locations in close proximity to the perceived untameability of the bush lends credence to Gelder and Weaver's analysis of the way in which the bush is utilised in colonial crime stories: "criminality saturating every level of bush life", suggesting that the bush itself somehow facilitates intra-human violence (2008, 6).

These texts therefore remain highly congruous with settler-colonial understandings of the Australian natural world and the nonhuman, problematically denying the role that such understandings have played, and continue to play, in the modern environmental crisis. Nevertheless, their exploration of the links between nonhuman and human communities, whilst representational to an extent, also reflects an important cause-and-effect relationship. As the climate crisis intensifies, nonhumans will continue to suffer and, in some cases, die alongside the humans affected; how this occurs in *Scrublands* and *The Dry* is therefore not simply symbolic of human suffering alone, but the very real connection between the suffering of humans and nonhumans in the context of the climate crisis.

References

Armstrong, Philip. 2008. *What Animals Mean in the Fiction of Modernity*. Milton Park: Routledge.

Ashcroft, Bill, Gareth Griffiths, and Helen Tiffin. 2002. *The Empire Writes Back: Theory and Practice in Post-Colonial Literatures*. 2nd ed. London and New York: Routledge.

Ashman, Nathan, ed. 2023. *The Routledge Handbook of Crime Fiction and Ecology*. New York: Routledge.

Bacigalupi, Paolo. 2009. *The Windup Girl*. San Francisco: Night Shade Books.

Bacigalupi, Paolo. 2015. *The Water Knife*. New York: Alfred A. Knopf.

Birch, Tony. 2017. Climate Change, Recognition and Social Place-Making. *Sydney Review of Books*, March 3. https://sydneyreviewofbooks.com/essay/climate-change-recognition-and-caring-for-country/. Accessed 29 Jan 2025.

Bohannon, Richard. 2014. Introduction. In *Religions and Environments: A Reader in Religion, Nature and Ecology*, ed. Richard Bohannon, 1–6. London and New York: Bloomsbury.

Brandi, Mark. 2017. *Wimmera*. Sydney: Hachette Australia.

Brandi, Mark. 2019. *The Rip*. Sydney: Hachette Australia.

Connolly, Robert. 2020. *The Dry*. Made Up Stories.

Cook, Kenneth. 1961. *Wake In Fright*. London: Michael Joseph.

Cordier, Stephane Christophe. 2018. Tim Winton's *In The Winter Dark* and the Settler Condition. *Antipode* 32 (1–2): 58–72.

CSIRO and Bureau of Meteorology. 2020. *State of the Climate 2020*. http://www.bom.gov.au/state-of-the-climate/documents/State-of-the-Climate-2020.pdf. Accessed 11 Feb 2025.

Dimick, Sarah. 2018. From Suspect to Species: Climate Crime in Antti Tuomainen's *The Healer*. *Mosaic: An Interdisciplinary Critical Journal* 51 (3): 19–35.

Donovan, Josephine. 2016. *The Aesthetics of Care: On the Literary Treatment of Animals*. New York: Bloomsbury.

Doyle, Briohny. 2016. *The Island Will Sink*. Melbourne: The Lifted Brow.

Driscoll, Catherine, Kate Darian-Smith, and David Nichols. 2017. *Cultural Sustainability in Rural Communities: Rethinking Australian Country Towns*. Abingdon and New York: Routledge.

Farrier, David. 2018. Animal Detectives and "Anthropocene Noir" in Chloe Hooper's *A Child's Book of True Crime. Textual Practice* 32 (5): 875–893.
Fetherston, Rachel. 2019. "No Longer Any Distance between Us": Technology, Climate Change and Narrative Empathy in Briohny Doyle's *The Island Will Sink. Antithesis Journal* 29: 36–45.
Fetherston, Rachel. 2023. Unsettlement, Climate, and Rural/Urban Place-Making in Australian Crime Fiction. In *The Routledge Handbook of Crime Fiction and Ecology*, ed. Nathan Ashman, 78–90. New York: Routledge.
Flannery, Tim. 1994. *The Future Eaters*. Port Melbourne: Reed Books.
Ford, Clementine. 2018. The Problem with the "Good Bloke" Narrative. *The Sydney Morning Herald*, May 14. https://www.smh.com.au/lifestyle/life-and-relationships/the-problem-with-the-good-bloke-narrative-20180514-p4zf6r.html. Accessed 30 Jan 2025.
Fox, Candice. 2017. *Crimson Lake*. Sydney: Bantam Press.
Gelder, Ken. 2001. Recovering Australian Popular Fiction: Towards the End of Australian Literature. In *Australian Literary Studies in the 21st Century*, ed. Philip Mead, 112–120. Association for the Study of Australian Literature.
Gelder, Ken, and Rachael Weaver. 2008. Colonial Australian Crime Fiction. In *The Anthology of Colonial Australian Crime Fiction*, ed. Ken Gelder and Rachael Weaver, 1–10. Carlton: Melbourne University Press.
Gibson, Ross. 2002. *Seven Versions of an Australian Badland*. St Lucia: University of Queensland Press.
Haddrick, Greg and Brenda Pam. 2008–2013. *Underbelly*. Australian Film Finance Corporation.
Hammer, Chris. 2010. *The River: A Journey Through the Murray-Darling Basin*. Carlton: Melbourne University Press.
Hammer, Chris. 2018. *Scrublands*. Crows Nest: Allen & Unwin.
Hammer, Chris. 2019. *Silver*. Sydney: Allen & Unwin.
Hammer, Chris. 2020. *Trust*. Sydney: Allen & Unwin.
Harper, Jane. 2016. *The Dry*. Sydney: Pan Macmillan Australia.
Harper, Jane. 2017a. *The Dry*. B Format Edition. Sydney: Pan Macmillan Australia.
Harper, Jane. 2017b. *Force of Nature*. Sydney: Pan Macmillan Australia.
Harper, Jane. 2018. *The Lost Man*. Sydney: Pan Macmillan Australia.
Harper, Jane. 2022. *Exiles*. Sydney: Pan Macmillan Australia.
Hensley, Christopher, John A. Browne, and Caleb E. Trentham. 2017. Exploring the Social and Emotional Context of Childhood Animal Cruelty and Its Potential Link to Adult Human Violence. *Psychology, Crime & Law* 24 (5): 489–499.
Hume, Fergus. 1886. *The Mystery of a Hansom Cab*. Melbourne: Kemp and Boyce.
Janson, Julie. 2022. *Madukka The River Serpent*. Crawley, WA: University of Western Australia Press.

Jones, Roger, Celeste Young, John Handmer, Adrian Keating, Gayathri Mekala, and Peter Sheehan. 2013. *Valuing Adaptation Under Rapid Change*. Gold Coast, QLD: National Climate Change Adaptation Research Facility.

Kearns, Laurel. 2014. Religion and Environmental Justice. In *Religions and Environments: A Reader in Religion, Nature and Ecology*, ed. Richard Bohannon, 297–311. London and New York: Bloomsbury.

King, Stewart. 2021. Crimate Fiction and the Environmental Imagination of Place. *The Journal of Popular Culture* 53 (6): 1235–1253.

Knight, Stephen. 1997. *Continent of Mystery: A Thematic History of Australian Crime Fiction*. Carlton South: Melbourne University Press.

Knight, Stephen. 2006. Crimes Domestic and Crimes Colonial: The Role of Crime Fiction in Developing Postcolonial Consciousness. In *Postcolonial Postmortems: Crime Fiction from a Transcultural Perspective*, ed. Christine Matzke and Susanne Mühleisen, 17–33. Amsterdam and New York: Rodopi.

Knight, Stephen. 2011. Peter Temple: Australian Crime Fiction on the World Stage. *Clues: A Journal of Detection* 29 (1): 71–81.

Langton, Marcia. 1996. What Do We Mean by Wilderness?: Wilderness and Terra Nullius in Australian Art [Address to The Sydney Institute on 12 October 1995]. *The Sydney Papers* 8 (1): 10–31.

Leane, Jeanine. 2010. Aboriginal Representation: Conflict or Dialogue in the Academy. *The Australian Journal of Indigenous Education* 39 (Supplement): 32–39.

Liu, Cixin. 2014. *The Three-Body Problem*, trans. Ken Liu. New York: Tor Books.

Louttit, Chris. 2015. The Mystery of a Hansom Cab, the Classic Adaptation and the Australian Canon. *Adaptation* 9 (1): 58–67.

Lucashenko, Melissa. 2018. *Too Much Lip*. St Lucia: University of Queensland Press.

Maslin, Janet. 2017. *The Dry*, a Page-Turner of a Mystery Set in Parched Australia. *The New York Times*, January 9.

Mayer, Sylvia. 2006. Literary Studies, Ecofeminism and Environmentalist Knowledge Production in the Humanities. In *Nature in Literary and Cultural Studies*, ed. Catrin Gersdorf and Sylvia Mayer, 111–130. Amsterdam and New York: Rodopi.

McLaren, Philip. 2002. *Scream Black Murder*. Philadelphia: Intrigue Press.

Media Watch. 2018. Reporting on Murder Suicides. *ABC News*, May 21. https://www.abc.net.au/mediawatch/episodes/reporting-on-murder-suicides/9972228. Accessed 30 Jan 2025.

Meixner, Sophie, and Tom Lowrey. 2020. Dead Livestock Paint Grim Picture of Fire Devastation and Logistical Challenges of Recovery Effort. *ABC News*, January 5. https://www.abc.net.au/news/2020-01-05/fire-bushfire-dead-livestock-farmers-agforce-animal-carcasses/11841868. Accessed 30 Jan 2025.

Mitchell, David. 2014. *The Bone Clocks*. London: Sceptre.

Morton, Timothy. 2016. *Dark Ecology: For a Logic of Future Coexistence*. New York: Columbia University Press.

Oerlemans, Onno. 2004. *Romanticism and the Materiality of Nature*. Toronto: University of Toronto Press.

Pollak, Michael, and Margaret MacNabb. 2000. *Hearts and Minds: Creative Australians and the Environment*. Alexandria: Hale & Iremonger.

Potter, Emily. 2014. Postcolonial Atmospheres: Recalling Our Shadow Places. In *By-Roads and Hidden Treasures: Mapping Cultural Assets in Regional Australia*, ed. Paul Ashton, Chris Gibson, and Ross Gibson, 75–86. Crawley: UWA Publishing.

Potter, Emily. 2019. *Writing Belonging at the millennium: Notes from the Field on Settler-Colonial Place*. Bristol: Intellect Books.

Renaldi, Erwin, and Erin Handley. 2020. Jakarta Residents Inundated as Severe Flooding Kills Dozens of People in Indonesia. *ABC News*, January 2. https://www.abc.net.au/news/2020-01-02/indonesia-jakarta-flooding-victims-indundated-heavy-rains/11837612. Accessed 30 Jan 2025.

Rigby, Kate. 2008. "Lifting Our Eyes from the Page": Religion, Literature and the Earth: Editorial Introduction. *Religion & Literature* 40 (1): 1–9.

Rigby, Kate. 2009. Writing in the Anthropocene: Idle Chatter or Ecoprophetic Witness? *Australian Humanities Review* 47 (Nov): 173–187.

Rose, Deborah Bird. 2013. Anthropocene Noir. *Arena Journal* 41:206–219.

Savery, Henry. 1830. *Quintus Servinton: A Tale, Founded upon Incidents of Real Occurrence*. Hobart: Henry Melville.

Scholtmeijer, Marian. 1993. *Animal Victims in Modern Fiction: From Sanctity to Sacrifice*. Toronto: University of Toronto Press.

Seet, Shawn. 2012. *The Mystery of a Hansom Cab*. Burberry Productions.

Simpson, Nardi. 2020. *Song of the Crocodile*. Sydney: Hachette Australia.

Sjö, Sofia. 2016. Madness, Otherness and Transformation: Exploring Religion in Nordic Crime Films. *Journal of Religion and Film* 20 (1): 1–21.

Smee, Ben. 2019. Brave or Naive? Conservative Wagga Surprises with Climate Emergency Declaration. *Guardian Australia*, July 20. https://www.theguardian.com/australia-news/2019/jul/20/brave-or-naive-conservative-wagga-surprises-with-climate-emergency-declaration. Accessed 30 Jan 2025.

Sussex, Lucy. 2015. *Blockbuster!: Fergus Hume and the Mystery of a Hansom Cab*. Melbourne: Text Publishing.

Tuomainen, Antti. 2013. *The Healer*, trans. Lola Rogers. New York: Henry Holt and Company.

Turnbull, Sue. 2023. Monstrous Wounds: Crime, Environmental Catastrophe and Domestic Abuse in Jane Harper's *The Dry*. *Journal of Australian Studies* 47 (2): 1–13.

Viskic, Emma. 2015. *Resurrection Bay*. Richmond: Echo.

Walter, James. 2020. Scott Morrison's Biggest Failure in the Bushfire Crisis: An Inability to Deliver Collective Action. *The Conversation*, January 8. https://theconversation.com/scott-morrisons-biggest-failure-in-the-bushfire-crisis-an-inability-to-deliver-collective-action-129437. Accessed Jan 30 2025.

Walton, Jo Lindsay, and Samantha Walton. 2018. Introduction to Green Letters: Crime Fiction and Ecology. *Green Letters* 22 (1): 2–6.

White, Lyn. 1996. The Historical Roots of Our Ecologic Crisis. In *The Ecocriticism Reader: Landmarks in Literary Ecology*, ed. Cheryl Glotfelty and Harold Fromm, 3–14. Athens, GA: University of Georgia Press.

White, Christian. 2018. *The Nowhere Child*. South Melbourne: Affirm Press.

Wilson, Laura. 2018. The Best Recent Crime Novels - Review Roundup. *The Guardian*, December 21. https://www.theguardian.com/books/2018/dec/21/best-recent-crime-novels-review-roundup. Accessed 30 Jan 2025.

Winton, Tim. 1988. *In the Winter Dark*. Carlton: McPhee Gribble.

Wisker, Gina. 2021. Shadows in Paradise: Australian Gothic. In *The Routledge Companion to Australian Literature*, ed. Jessica Gildersleeve, 384–392. New York and London: Routledge.

Yell, Susan. 2010. "Breakfast Is Now Tea, Toast and Tissues": Affect and the Media Coverage of Bushfires. *Media International Australia, Incorporating Culture & Policy* 137: 109–119.

Zhang, Yang, Samsung Lim, and Jason John Sharples. 2017. Effects of Climate on the Size of Wildfires in the *Eucalyptys camaldulensis* Forests and the Dry Lands of the Riverina Biogregion, Australia. *Forest Ecology and Management* 401: 330–340.

CHAPTER 7

Historical Ecofiction and the Agency of Water

7.1 Shifting Water Rights and the (Post)Colony

When we talk about ecology, no matter the place or time, we are also inherently talking about the seemingly elusive forms of air and water. It is perhaps appropriate, then, to turn to the agency of water in this final chapter. The final text I would like to consider here is Wiradyuri author Anita Heiss' *Bila Yarrudhanggalangdhuray River of Dreams* (2021). I approach this novel as a work of historical ecofiction that asserts Indigenous sovereignty and belonging through its depiction of water agency, evoking the idea that histories are constantly shifting and, in the case of colonialism, often warrant rewriting. Heiss' novel subverts the traditional form of the colonial historical novel through her use of the Wiradyuri language alongside English and her portrayal of colonial dispossession as ongoing rather than an event confined to the past. The novel's focus on the agency of water and rivers reveals the instability of settler belonging, both now and in the past, the narrative beginning with the historical and catastrophic flooding of Gundagai in 1852—a town built by settlers on a floodplain.

I would like to frame my approach to water in two contexts here—one in which I consider the materiality and agency of water in the broader context of the Australian historical fiction genre, and another in which I look closely at water as it is represented relationally and agentially in Heiss' novel. Before I discuss Heiss' work, though, I would like to briefly attend

R. Fetherston, *Theorising the Postcolonial Eco-Novel*, Literatures, Cultures, and the Environment,
https://doi.org/10.1007/978-3-032-04466-2_7

to the importance of shifting human–water relations globally and the unique ways in which historical fiction subverts dominant colonial narratives and histories.

My purpose here is not to downplay or dismiss the destructive potential of water. At the time of writing parts of this particular chapter, Hurricane Helene made landfall in the south-eastern states of the United States. Some of the research for other chapters of this book was undertaken in early 2020 as extreme flooding decimated parts of Indonesia. I am therefore cognisant of the increasingly destructive and tragic impacts that water is having on populations across the world due to anthropogenic climate change—at least as cognisant as a person who is geographically and socially removed from the impacts of these disasters can be. Rather than purely a "force of nature", though, I want to approach water as agential, particularly in the way that it shapes human and nonhuman lives while also being shaped itself, and how this agency is represented in Australian historical fiction.

Key to this chapter's discussion is Heiss' positioning as an author of genre fiction—in this case, historical fiction. In simplistic terms, we might consider historical fiction as consisting of fictional narratives set in a previous era to the era in which the narrative itself is produced. Kate Grenville's *The Secret River*, for example, was published in 2005, but set in the early nineteenth century and tells the story of a fictional convict-turned-freeman William Thornhill settling along the Hawkesbury River in New South Wales. Historian Tom Griffiths describes the form of the realist historical novel as emerging from within the colonial era itself and thus often reflecting the deterministic, individualist ideologies of the time (2015, 16). Nicholas Birns argues something similar when he describes how the Australian historical novel emerged from "late eighteenth- and nineteenth-century Australian exploration, colonization, and settlement"—a literary form which, he goes on to say, "has dominated Australian literature for over a generation" (2018, 7). Despite its origins in colonialism, the Australian historical novel has, according to Birns, "played an important role in rebalancing Australian historiography and culture towards a greater awareness of the ravages of settlement and the durability of the anterior and posterior Indigenous presence of the land" (2018, 7).

This is particularly true when we consider the heightened popularity of the settler-authored historical fiction novel in the early 2000s in Australia. This includes the aforementioned *The Secret River* by Grenville, Peter Carey's *True History of the Kelly Gang* (2000), Colleen McCullough's

Morgan's Run (2000) and *The Touch* (2003), and Richard Flanagan's *Gould's Book of Fish* (2001). These novels reflect the political atmosphere in Australia at the time which has come to be known as the "history wars": a period in which various historians, politicians and cultural commentators were debating the true extent of colonial violence and the level of responsibility the Australian government and other authoritative bodies should or should not be taking to make reparations to First Nations people.

Many of the historical novels that emerged at this time, including *The Secret River*, have been referred to as "sorry novels" (Weaver-Hightower 2010). Such novels arguably did important work highlighting the violent, often untold histories of colonial Australia by engaging with themes of "non-indigenous refusal, denial, guilt and responsibility" (Potter 2019, 46). However, this kind of historical novel is also what Emily Potter refers to as a means for "non-indigenous Australians to redeem their fraught belonging and reclaim a rightful place" (2019, 48). Authored by non-Indigenous writers and usually featuring a white, non-Indigenous protagonist, the sorry novels are largely depictions of colonisation from a non-Indigenous, settler perspective. As Potter also writes, these narratives "prioritize the effects of historical recognition and legal redress … on non-indigenous subjectivity as they simultaneously acknowledge … a reckoning with the same history" (2019, 48).

Returning to Grenville's *The Secret River*, Griffiths writes that Grenville's primary aim in this historical novel is "to dramatize the violence—so that the reader may experience it" (2015, 15). What the reader experiences, though, is this violence through a primarily non-Indigenous perspective, despite the most extreme and deplorable violence usually occurring to Indigenous characters in the novel. First Nations works of historical fiction, such as Heiss' novel, are therefore not just responding and providing counter-narratives to historical and political debates regarding colonisation—they are also challenging the white settler voice that remains dominant in a vast number of Australian historical novels.

Historical fiction also occupies an unusual niche in comparison to other genre fictions. This is because it is often considered separately from the more "popular" genre forms such as crime fiction, science fiction and fantasy fiction. Historical fiction seems to tread the line between the popular and the literary more closely than other genres. Australian novels such *The Secret River*, Hannah Kent's *Burial Rites* (2013), Melissa Lucashenko's *Edenglassie* (2023), Mirandi Riwoe's *Stone Sky Gold Mountain* (2020) and Richard Flanagan's *The Narrow Road to the Deep North* (2013) may technically be works of historical fiction, but they have also been considered within the broader classification of "Australian literary fiction". Such novels have won awards and been acclaimed by critics in terms that identify them as more than historical genre fiction, and as works that are literary in nature despite their clear adherence to the form of the historical novel.

Heiss' *Bila Yarrudhanggalangdhuray* is no different, having been awarded the 2022 New South Wales Premier's Literary Awards' Indigenous Writers' Prize, highly commended in the 2022 Victorian Premier's Literary Awards and longlisted for the 2022 Stella Prize—all considered some of Australia's most prestigious literary awards. Significantly, some reviews of Heiss' work reveal its simplicity and broad appeal, which returns us to its categorisation as genre fiction. Writing in *Guardian Australia*, Beejay Silcox describes *Bila Yarrudhanggalangdhuray* as "a simple story, simply told" (2021), while Jane Sullivan in *Australian Book Review* asserts that the novel's "strong emotional pull and … accessible female hero" makes the work deserving of "wide appeal" (2021). While I do not have the space in this chapter to explore the complexities of literary awards and review culture and what they may or may not denote about a novel, I believe the above provides some important context for the key point being made here—that historical fiction, whether considered literary "high-brow" or not, is engaging in diverse and yet accessible ways with the more-than-human and environmental crisis.

Unlike other genre fiction, historical fiction does not necessarily rely on a repetitive formula, depending primarily on "the formal techniques and cultural assumptions of the main traditions of the novel", as Harry E. Shaw describes (1983, 23). It is therefore a genre that "does not have a significant history apart from the history of the novel as a whole" (Shaw 1983, 23). The category of *Australian* historical fiction is complicated here by differing conceptions of time between settler-Australian cultures and Indigenous cultures. The past decade of Australian publishing has demonstrated increased attentiveness to First Nations historical fiction—particularly novels that reframe and challenge colonial accounts of the British Empire's invasion and continued colonisation of the continent. While Heiss' novel has been marketed by the Australian publishing industry as a work of historical fiction, it is also a piece of First Nations literature that reflects Aboriginal understandings of history and time that differ considerably to the Western emphasis on linear time and "progress" from past to present to future, as a work like *The Secret River* is particularly concerned with.

Tara June Winch's award-winning novel *The Yield* (2019), for example, tells three interweaved narratives, two of which are set in the present day depicting the ongoing colonisation of Wiradyuri land via the mining industry, and the third set in the past, told through the letters of a white missionary working in this same area in the early twentieth century. As one reviewer remarked, "*The Yield* works to reclaim a history that never should have been lost in the first place" (Wagner 2021)—a comment which could indeed be applied to many works of First Nations historical fiction, including Heiss' *Bila Yarrudhanggalangdhuray*. These novels do more than this, though, often subverting reader expectations of the historical fiction form which has largely been dominated by white, settler-Australian authors. A recurring focus on water across many of these Indigenous Australian novels is what I am especially interested in discussing in this chapter.

In the humanities, scholarly engagements with water—often referred to as the "blue humanities"—have largely been concerned with saltwater environs, primarily seas and oceans. Turning inland, to the waters of Australia's interior, is an opportunity to reframe some blue humanities scholarship. In *Bodies of Water*, Astrida Neimanis considers the human body as itself a body of water—not simply because the human body is "made mostly of wet matter" but because we are also "aswim in the discursive flocculations of embodiment as an idea" (2016, 1). It is not possible to think through the complexities and messiness of water relations in the Anthropocene without also considering the human as itself a site of watery embodiment.

This idea is not necessarily new to many First Nations peoples, but, as Neimanis states, it *is* new in the context of Western humanist perspectives on embodiment, which frame bodies "as discrete and coherent individual subjects" (2016, 2). Significantly, the idea—or material reality—of bodies as water actually challenges the notion that bodies belong solely to the human (Neimanis 2016, 2). As Neimanis describes, "As bodies of water we leak and seethe, our borders always vulnerable to rupture and renegotiation" (2016, 2). Taking texts within Australian ecofiction and trying to understand what they have to say about human–water relations is, in my mind, one step in this renegotiation process. How are writers representing water and humans' simultaneous connection with and distance from it, and what can we learn from this? This emphasis on water is particularly pertinent, pointing clearly to climate crisis in so many ways: rising seawaters, ocean temperatures, coral bleaching, flash flooding, drought, bushfire and resource availability Water, including human–water relations, is central to all of these things and, as Neimanis also writes, it "constitutes one of the so-called Anthropocene's most urgent, visceral, and ethically fraught sites of political praxis and theoretical inquiry" (2016, 20).

Water is often deeply tied to human memory, and so looking at its representation in an historical fiction context makes particular sense. John Jerome writes that "water has a way of giving memory back. For some of us water is memory, and memory water" (1997, 8, cited in Thrush 2017, 272). If this is the case, then it logically follows that the devastation and disappearance of bodies of water in the Anthropocene can result in amnesia, according to Coll Thrush (2017, 272). Using the example of the diversion of the White River in Auburn, Washington, Thrush describes his experience of attending a school built where the river once flowed. "Never once", he writes, "had we heard that a river flowed under these classrooms and recess yards, even as we built our own memories in its path" (2017, 272).

To bring this point back to the Australian context, we might apply Australian First Nations author and historian Tony Birch's experience of observing the Birrarung (Yarra River) in the eastern suburbs of Melbourne as it was drastically altered by rapidly expanding road infrastructure. Returning to a particularly special part of the river he had once visited as a child, Birch recalls the moment he recognised the far-reaching impacts of urbanisation on his own connection to place: "The Eastern Freeway runs along the southern fringe of the billabong, and the flow of traffic is so constant that sitting by the water and listening, it is difficult to discern the sounds of an individual car or truck engine" (2018). At the time of my own writing, this area of Melbourne has been undergoing further significant road construction to alleviate the traffic congestion in this part of the city—congestion that the Eastern Freeway and other existing roads can no longer withstand. Many of the parks, vegetation and walking paths that were created during the earlier road construction have now been destroyed, and these new impacts on the river and its associated ecosystems are only just being seen. Birch writes of the importance of expressing love for places like the Birrarung through story, as well as listening to First Nations stories about these places (2018).

For humans and many other animals, water enables mobility, hydration, sustenance, habitat, temperature regulation and so much more. Rain, including its absence, is one of the most significant ways in which humans encounter water. As Mick Smith describes, rain—particularly the feeling of it falling onto one's skin—draws attention to our material embodiment and can lead to further recognition of how one sits in relation to other lives: "To feel our exposure to rains falling and not forget to give thanks is to partake in events precipitating around us, to be ecologically open to life, despite our eventual mortality" (2017, 230–231).

Water is floated on, swum through, drunk, harvested and diverted, and its movement used as a form of energy in various contexts. It is easy to see, then, how water has long been considered more object than life in Western culture—it is often construed as a medium or resource rather than a more-than-human entity in its own right. It is something to be utilised, something that sustains other life that relies on it, such as the human, but not a life in itself. As Smith describes in reference to Martin Heidegger's writings on technology, the failure to properly appreciate something such as rain and one's inextricable relation to it is a result of the "thoughtless appropriation of every 'thing' as just another potentially transformable *object*, a resource passively awaiting human instrumental purposes" (2017, 232). Within this framework, rain is "just a source of water for agricultural production, a supply of potable liquid, a matter of environmental security, a coolant, solvent, and … source of hydroelectric power" (Heidegger 1993, cited in Smith 2017, 232). This instrumentalist reduction of rain and water is particularly significant in the context of Australia—a drought-prone continent where settler-colonial practices of water diversion, theft and pollution have left an indelible mark on lands, waterways and human and nonhuman inhabitants. There are, of course, many non-Western cultures and ontologies which approach water differently, such as the Wiradyuri people whose connection with water and rivers is explored in Heiss' novel. Understanding Western-colonial approaches to water, though, is key to understanding the origins and current trajectory of the modern ecological crisis, as well as how Heiss' novel subverts such approaches through the historical novel form.

Water as an aesthetic and narrative device in literature has an equally complex history. Even in its perceived absence, it is there—both in our physical surroundings and in literature. It is also important to consider more than just literary representations of the surface waters that are often the main way in which water is represented in literature (i.e., rivers, lakes, seas and oceans). For example, groundwater—"waterflows in underground aquifers where the interstitial spaces in solid rock and soil are filled or saturated with water" (Wardle 2024, 2)—is a key player across ecosystems and yet, as Deborah Wardle contends, "fictional and poetic impressions of groundwater are rare" because "it is hard to see and difficult to imagine" (2024, 3). While this chapter is not especially concerned with groundwater (although it is no doubt present beneath the environs that Heiss depicts in her novel), a similar notion applies to how water is considered across Western literature more generally. It is certainly *there*, as part of a narrative's setting or used as metaphor for the changes, cleansing or struggles that occur to (usually human) characters. It is certainly *there* but it is rarely remarked upon, at least in Western literature, as a character in its own right that influences events of the novel and has things happen to it, as well as making things happen to others. Similarly, the *agency* of water—something that is hard for many in the West to see and imagine—is largely missing from much Australian settler literature. First Nations historical fiction, such as Heiss' work, provides more agential depictions of water. In this chapter's analysis, I attempt to position water as more than a backdrop, more than allegorical, more than metaphor, and rather as a material agent with stories of its own.

Water is a complex figure in Australian literary traditions, particularly given the stereotype of the continent as a "wide brown land", defined by drought and aridity, as depicted in Dorothea Mackellar's poem "My Country" (1908). There is a certain irony surrounding the fact that one of Australia's other most well-known environments is the Great Barrier Reef. This wet/dry dichotomy follows through to Australian literary and cultural traditions, particularly those of the setter-coloniser. Grenville's *The Secret River* is again a notable example here—a novel that designated the millennial shift in Australian settler writings towards more complex considerations of settler–Indigenous relations, while also exemplifying an indelible attachment to certain colonial literary traditions that position Indigenous people and the environment as oppositional to settler ways of being and living on the land (and water). Grenville positions a river (in this case, the Hawkesbury River, located north-west of Sydney) as central to the settler struggle for belonging in early colonial times, but it is still largely figured as a backdrop to the settler narrative. What I assert in this chapter is that understanding Australian ecologies as agential but also ambiguous in their wetness or dryness provides a more relational approach that makes space for messier considerations of human–nature relations—considerations that resist dichotomous narratives of supposedly oppositional environments, such as "wet" versus "dry". Borrowing Donna Haraway's term, we must "stay with the trouble" (2016) of amorphous water if we are to understand how to live as part of and alongside it—through floods, droughts, storms, pollution and shifting shorelines.

We might also identify certain narrative techniques used in Australian settler literature that reveal particular attitudes towards water. Teresa Shewry and Philip Steer recognise the use of irony across various works of Australian settler literature as a means of reflecting "water's resistance or excessiveness to settler knowledges" (2024, 353). In this case, irony is well positioned to reveal the "doubleness" of settler approaches to water relations in both colonial and contemporary Australia. Representations of situational irony in both older and more recent forms of Australian settler writings "simultaneously offer a critique of water's failure to meet expectations and signal the possibility of alternative understandings" (Shewry and Steer 2024, 354).

First Nations literature in Australia is also deeply concerned with water, but often in vastly different ways to Australian settler literature. Julie Janson's *Madukka The River Serpent* (2022) is a recent example. A work of Aboriginal crime fiction, Janson's novel explores the impact of colonial environmental policy and law enforcement on Aboriginal communities and the health of inland waterways. The river is an agential figure in this text, an enactor of justice for First Nations people and the environment, and the material embodiment of Aboriginal sovereignty and cultural strength (Fetherston 2023). Similarly, Melissa Lucashenko's *Too Much Lip* (2018) depicts a river as a material, agential force—a more-than-human being that speaks for Country and plays a key role in the protagonist's fight to protect the land of her family and ancestors from the development of a new prison. Kathrin Bartha-Mitchell argues that it is in fact the fluid, mobilised "experiences of displacement, ostracism and hybridity" across both rural and metropolitan places in twenty-first-century Australia that allow the Indigenous characters of Lucashenko's novel to solidify their custodianship of the river (2024, 173). This is represented by the mobility of the river itself and the "animals, ancestors [and] elements" connected to it (Bartha-Mitchell 2024, 173).

Neither of these texts is classified as historical fiction, though; while significant examples of Indigenous Australian representations of water and rivers in literature, they are set in the here and now. As I will go on to describe, Heiss similarly emphasises water as agential but instead writes into the Australian historical fiction genre while concurrently subverting some key tropes of it. What unites the above texts with a work like Heiss' novel is their interest in connecting waters and rivers with the fluidity of time—not in a metaphoric sense, but in a very real, material sense in terms of how rivers convey the agency of Country and the continued sovereignty of First Nations people, past, present and future, over their traditional lands, waters and skies.

7.2 Anita Heiss' *Bila Yarrudhanggalangdhuray* and Waterway Agency

Heiss' 2021 novel *Bila Yarrudhanggalangdhuray* is a work of historical fiction that positions water as a central figure in Wiradyuri and colonial Australian history. The novel begins in 1838, the prologue describing the early construction of the town of Gundagai on the flat land close to the Murrumbidgee River—usually referred to throughout the narrative as the Marrambidya or "the bila". Following the prologue, the narrative shifts to the day of "The Great Flood" in Gundagai on 24 June 1852 (Heiss 2022, 13) and from there follows the experiences of young Wiradyuri woman Wagadhaany as she is forced to travel to another town, Wagga Wagga, as an indentured servant to a settler family, the Bradleys—two brothers who lost their parents and a brother in the Gundagai flood. Wagadhaany develops a close relationship with Louisa, the new wife of the eldest Bradley brother, James, and is a close confidant of hers as Louisa deals with her husband's ambitions to become mayor of Wagga Wagga and his worsening alcoholism and violence. Louisa is a complicated figure—a Quaker who sees herself as well-meaning in her efforts to help Aboriginal people, particularly women, but, as the reader learns through Wagadhaany's eyes, is more of a white saviour who oppresses Wagadhaany and understands little about Wiradyuri culture. The novel ends five years after the Bradleys move to Wagga Wagga, with Wagadhaany, her husband Yindy and their children escaping their servitude there and arriving at the mission where Wagadhaany's family have been moved to since she last saw them in Gundagai.

While the majority of the novel's action takes place in Wagga Wagga after Wagadhaany is forcefully relocated there, it is worth first outlining a brief history of the town of Gundagai, particularly its environmental history. First surveyed in 1838 by the New South Wales colonial government, Gundagai was established in 1841 as a key stopover town for people travelling overland from Sydney to Melbourne (or Port Phillip) and South Australia (O'Gorman 2012, 15–16). Significantly, it was the only place at the time where travellers could cross the Murrumbidgee River (O'Gorman 2012, 15). Today, the 1852 Gundagai flood is still considered Australia's deadliest flood in its recorded colonial history (NMA). This flood is famous not just because of the number of lives lost—estimates range from 75 to 100 or, to put this in perspective, one-third to one-half of the people living in or visiting Gundagai at the time (O'Gorman 2012, 16)—but because, as a result of the destruction, large parts of the town had to be entirely relocated to an area north of the original town known as Mount Parnassus. It is appropriately described by environmental historian Emily O'Gorman as "the town that moved" (2012, 15) and if you were to visit the town of Gundagai today, this "newer" location is where you would find it.

O'Gorman argues that the 1852 flood in Gundagai has played a vital role in understanding and constructing flood histories in Australia, particularly the way environmental knowledge has been "accumulated, contested, and evaluated" in the context of both local and broader colonial systems (2012, 17). Heiss' depiction of this historical event also does this, but in a way that not only draws attention to environmental knowledge associated with this period, but to the narrative traditions embedded in the Australian historical novel that both facilitate and limit different histories that can be told.

The Wiradyuri people are the traditional owners and custodians of the land on which the township of Gundagai sits; "Wiradyuri" means "people of the three rivers" (O'Gorman 2012, 17). The establishment of the town of Gundagai occurred in the context of increased colonial violence in the area, sometimes referred to as the "Wiradjuri War" (Gammage 1983, cited in O'Gorman 2012, 20). It is also worth noting that the more common spelling of this First Nations group in contemporary Australia is "Wiradjuri". However, Heiss makes the point in her "Glossary of Wiradyuri Words" that "there is no 'j' in our language alphabet. The 'j' sound is made by the letters 'dy'" (Heiss 2022, 377). She therefore chose to use the spelling "Wiradyuri" throughout her novel. I have also chosen to use this spelling in acknowledgement of this.

Yarri, Wagadhaany's father in the novel, as well as the characters Jacky Jacky and Long Jimmy, two other Wiradyuri men who rescue people from the flood, are all based on real people who did indeed rescue many Gundagai citizens from the 1852 flood. Heiss includes an important historical note at the end of her novel that explains this, describing how it took years for any of these men to receive rightful recognition for their heroism. Two of these men were eventually acknowledged in 1875 when Yarri and Jacky Jacky were "given an engraved breastplate each and granted a lifelong pension from the settlers" (Heiss 2022, 282). On the 165th anniversary of the flood in 2017, the town of Gundagai erected a sculpture in their honour, which "brought national attention to the men" (Heiss 2022, 383).[1]

In an article that describes the role of the real-life Yarri, or Yarrie, who rescued settlers from the flood in his bark canoe, Wardiningsih Soerjohardjo comments on the hubris of the white Gundagai residents who had indeed received "sufficient warning" in the form of the previous flood of 1844 (2012, 121). In fact, there had been multiple flooding events in the preceding years—one in 1844, 1847 and 1851 (O'Gorman 2012, 15). As Soerjohardjo goes on to say:

> Memories of the past, particularly of the 1852 flood and Yarrie's deeds, awakened by individuals—the flood's survivors and the witnesses—were then interwoven with each other, and they gradually accumulated into an imagination, and were finally shared amongst Gundagai citizens until they formally and explicitly became a received social expression. (2012, 123)

This is a memory that deflects from the continued dispossession and oppression of the local Wiradyuri people, most likely including the heroic Yarri and his family. Heiss depicts this in her novel, portraying Yarri's heroism during the floods when he rescues his daughter Wagadhaany, the subsequent award given to him by the townspeople, and, at the end of the novel, the removal of him and his family from Country to a mission in a different place. The dominant "social expression" that conveys the settlers' shared memories of the floods, while not entirely invalid, significantly displace the memories of Yarri himself and the other Indigenous people who experienced the flood, and colonial dispossession, first-hand.

Similar to historical accounts by Australian settlers, Australian settler literature has a reputation of making invisible the experiences and stories of First Nations people or otherwise depicting them in stereotyped, racist

[1] This same historical note also mentions a report that Yarri allegedly murdered a 16-year-old Aboriginal woman, Sally McLeod, in the same year of the flood. There was never an official conviction, though, Heiss commenting that "we don't know if Yarri did or didn't commit this heinous crime" and that she "chose to create a fictional Yarri whose role in the rescue could allow him to be a symbol of extraordinary and largely unacknowledged heroism at the time" (Heiss 2022, 384).

terms. Several poems by settler authors have been written about the flood and, while some acknowledge Yarri as a hero, others convey racist assertions about Aboriginal people: "a black man, but his heart was white", states an anonymously authored poem published in *The Gundagai Independent* in 1952 (cited in Soerjohardjo 2012, 128). Some of these verses primarily emphasise the experiences of white settlers during the flood and, unsurprisingly for the time, many neglect to mention the role of Aboriginal rescuers such as Yarri.

Heiss' novel represents a direct response to both historical accounts and literary accounts of the Gundagai flood by emphasising Wiradyuri experiences of the time as well as Wiradyuri cultural knowledge of rivers and water. Although this work is fictional, as a Wiradyuri woman herself Heiss is arguably presenting her own real-life reflection on the event given her cultural and familial ties to those Wiradyuri people who would have experienced it first-hand in 1852. In doing so, Heiss establishes the importance of human–water connection in Wiradyuri culture which in turn asserts Indigenous sovereignty. The accompanying "Glossary of Wiradyuri Words" (Heiss 2022, 377–381) included at the end of the book attests to the fact that this novel is not just a piece of historical fiction, but a work that seeks to re-affirm sovereignty of Country through language as well as educate non-Indigenous readers about Wiradyuri language and culture.

It is difficult to separate the timing of the novel's publication in 2021 from the contemporaneous water politics of the Riverina region—an area which includes the Murrumbidgee River of Heiss' novel. The Murrumbidgee is also considered a part of the Murray-Darling Basin, the management of which has come under increased scrutiny in the past decade because of "allegations of corruption and water theft, because of dead fish and angry irrigators, and because a royal commission in South Australia has suggested one of our important government organisations, the Murray–Darling Basin Authority, is dishonest, incompetent and acting outside the law" (Simons 2020, 3). This is just one example of many that point towards water as "unavoidably political" (Simons 2020, 3) in the dry country that is Australia. As Margaret Simons states, "water connects people, but it also divides" (2020, 2). While these modern water politics may seem separated from the events of Heiss' novel by over 150 years, the settlers' control of the Murumbidgee in the novel is arguably an early example of colonial water theft—an example that has set the stage for more contemporary cases of environmental catastrophe and continued Indigenous dispossession in the Riverina region. Heiss' novel was also a timely reminder of the power of floods in inland Australia, particularly given the 2021 publication of the novel coinciding with the extreme floods across significant parts of eastern Australia in late 2021 and early 2022.

The aforementioned *The Secret River* is arguably the quintessential Australian historical novel being subverted by Heiss here. In comparison to Alexis Wright's Aboriginal realism, Heiss' work does not engage in what (Western) readers would read as the "speculative"—it is largely realist, depicting "events that could have occurred and people portrayed could have lived" (Groce and Groce 2005, 205). In that sense, Heiss' work may seem similar to Grenville's approach to historical fiction. However, as I will go on to discuss, Heiss subverts the dominant narratives of Australian settler historical fiction via an increased focus on historical knowledge and accounts that exist outside of the settler archives.

To the white settlers in Heiss' novel, water is considered primarily a resource or a monstrous bringer of disaster. The settler inhabitants of Gundagai are depicted as lacking a fundamental understanding of water systems in inland Australia as ephemeral—flowing in one season, and dried up the next. This is knowledge that the Wiradyuri people, particularly Yarri, try their best to convey to the settlers as they begin building the town on the river flats: "'Flood area, biggest rains, the water goes past here.' He raises his hand up above his head to show how high the water may rise" (Heiss 2022, 4). In response, one settler tells Yarri, "You don't know what you're talking about. We *need* rain. It certainly doesn't look like it's going to flood to me" (Heiss 2022, 4). This conversation reflects historical knowledge of settlers' approaches to building Gundagai, and also the choice of many of them to remain in the town despite recent floods in the lead-up to 1852. O'Gorman describes how many settlers living in the area mistakenly assumed that significant floods were unusual—"after experiencing a large flood they believed that the 'big one' was behind them" (2012, 17). It was this belief, she writes, that is partly to blame for their choice to continue living on the river flat after the flooding events of previous years.

We see this belief illustrated in Heiss' novel when, following the devastating consequences of the flood, settlers are "reliving the history of previous floods and talking about how it should not have happened again" (2022, 51). These same settlers in the novel are also quick to blame government officials for "allowing the town to be flooded, for not being better prepared" (Heiss 2022, 51). This is certainly representative of some of the local and national media responses to the floods at the time. In the wake of the catastrophic 1852 flood, many townspeople blamed the government surveyors for the devastation—they had, after all, recommended the area to build on back in 1838. In reality, the government had actually surveyed and auctioned off nearby higher ground in response to the 1844 flood (Soerjohardjo 1986). Despite this, Soerjohardjo describes how "the people still preferred the lower ground, the bed of a river, even though their lives were in peril" (1986, 56, cited in O'Gorman 2012, 17).

Wiradyuri knowledge of the river is passed on to the younger Wagadhaany, both through her own direct experience of the waters and via the sharing of cultural knowledge by her uncles:

> She has heard her uncles talking about not building on the river flats, that it's not the right place to make a home for anyone. That it's not safe to a build a house where the water will flood when it rains heavily.... She knows this place has flooded before, and it will again. (Heiss 2022, 2)

The Wiradyuri characters certainly view water and the river in less dualistic ways than the settlers. This is perhaps made most clear when Wagadhaany's only daughter, Miima, drowns in the river while playing with her brothers during their journey from Wagga Wagga back to Gundagai in the final chapters of the novel. Wagadhaany tragically reflects, "The same bila that gave her so much joy with Yindy has devastated her. She wants to walk into the Marrambidya right now and not walk out, to stay there until her spirit meets her daughter's" (Heiss 2022, 367). Wagadhaany notes that her grief is not dissimilar to that experienced by the settlers who lost their families in the flood: "For a fleeting moment she feels for the Bradley brothers. Now she knows what it is like to have lost someone to the force of the Marrambidya" (Heiss 2022, 367).

What is different about Wagadhaany's experience of loss is the complexity of approaching grief with the knowledge that the river—a part of Country—is where her daughter died. Something that comforts Wagadhaany is the understanding that, as her mother once told her, "death means going back" (Heiss 2022, 367). Wagadhaany and Yindy bury Miimi by the river, sending her "to the Wiradyuri Ancestors who will be waiting for her" (Heiss 2022, 367). While a very different geographical and cultural context, Potawatomi writer Robin Wall Kimmerer's thoughts on rain and water bear relevance here. Describing her walk in rain along the Lookout Creek in Oregon, Kimmerer acknowledges the difficulty of really knowing a river, even if one is to recognise more-than-human agency and relationality. "It is a different river in August than in October", Kimmerer writes, "You'd have to stand here a long time to know them both. And even longer to know the river that was here before the coming of the gravel bar, and the river that will be after it leaves" (2020, 298). To truly *know* water is arguably impossible—in both Western and non-Western contexts. Given its fluid nature, its ability to seep into and out of everything, here one day and gone the next, how can one claim to know water intimately, even if as a part of ourselves? When Miima dies, Heiss' novel does not label the river as monstrous or callous, as the settlers do. Wagadhaany's response instead conveys both a knowability and unknowability about the river that allows it agency beyond both Wiradyuri and colonial knowledge systems.

While Miima's death is no less tragic than the deaths of the settlers during the flood, the settlers' grief and their attitudes towards the river following the flood are markedly different to Wagadhaany's. While attending the local church with the Bradley family after the flood, Wagadhaany notes that the priest's words suggest that the disaster "was the crooked work of God" (Heiss 2022, 55). Following the service, she is shocked to hear David Bradley say that he "got some answers today" (Heiss 2022, 55). Commenting later about their house's withstanding of the flood, David says that the river had tried "its darndest to tear it down" but that, thankfully, God had been "looking over us" (Heiss 2022, 57). Reflecting the settler notion, discussed in Chap. 6, of a vengeful Australian wilderness that is only granted some kind of agency by the coloniser when it wreaks havoc upon them, the settlers in Heiss' novel are quick to attribute an angry, violent personality to the river. Any other kind of agency regarding the river is rarely remarked upon by the settler characters in the novel.

Compared to the Wiradyuri characters, the stubbornness of the settlers surrounding their knowledge of the river is palpable and frustrating—they evince an assured knowability which they believe exceeds the ontologies and epistemologies of Wiradyuri culture that the settlers are not willing to learn from, even when offered such knowledge. Heiss draws the reader's attention to this, revealing the inadequacies of settler attitudes towards water and more-than-human life-worlds that not only affect the Indigenous people who are violently dispossessed of their land and waters, but the colonisers themselves as their settlements and livelihoods succumb to the force of the river. As Wagadhaany observes,

> She wonders if the White people understand that most Wiradyuri men know how to work with the river and the land, even in times of flood. That the Marrambidya is not something to be afraid of. Rather, the bila is to be respected and relied upon for food, for transport, for life. That the men have been brave *and* smart. (Heiss 2022, 62)

When it appears that the settlers' perspectives on water and the river have not drastically changed since the flood, Wagadhaany attempts to educate an interested Louisa about the Wiradyuri relationship with water:

> "Rain is life, Louisa"…. "It's simple. The earth gives us rain for water to drink, and food from the land to eat. This is how she shows she loves us. In return we must always show respect. If we respect the earth and sky, they will respect us back." (Heiss 2022, 137)

It is in interactions like these that Heiss demonstrates the ability of human–water relations to affect human–human relations as well. As Jeffrey Jerome Cohen argues in his analysis of different flood narratives, "The flood makes evident a lack of affective connection already present, an everyday inability of sympathy to cross boundaries of nation, race, species, class" (2017, 249). As he rescues a white man from the flood in the novel, whose clothes had been ripped off by the force of the water, Yarri considers the difference between himself and this man and the connection made possible through natural disaster and the more-than-human:

> … as the canoe floats with effort to the shore, Yarri thinks about the two men there together; a naked White man and a barely clothed Black man are nothing but two men stuck in the midst of a devastating flood. He knows he has the power to save this man's life. *A life is a life*, he says over and over in his mind, knowing that the weather, the rain and the river don't care what colour anyone is right now, and that in this moment they are equal. (Heiss 2022, 33)

Despite the empathy and heroism displayed by Yarri, Heiss also depicts the drastic impact of the flood on the local Wiradyuri people—as much as the flood reveals their knowledge of and connection to Country, it also leads to increased oppression and the expansion of colonial power over the surrounding land. Not long after the flood has occurred in the novel, there is already news circulating "that the townsfolk are going to petition the government to swap their land for lots on higher ground" (Heiss

2022, 51). Concerningly, though, Wagadhaany notes that "there is no mention of what will happen to her family's camp" (Heiss 2022, 51) that is already established in that location. She also recognises that "… while everything has changed for the townsfolk, she feels like absolutely nothing has changed for her…. She resents still being spoken to as the servant, the cleaner, the cook" (Heiss 2022, 49). To return to the previously quoted Simons, "water connects people, but it also divides" (2020, 2).

If one was to compare Heiss' historical novel to Wright's genre-bending *The Swan Book* (2013), it would be easy to recognise a difference in how each author approaches such divisions between Indigenous and non-Indigenous people. Wright's work certainly identifies the oppressed from the oppressor, but muddies the waters with characters like Warren Finch and Aunty Bella Donna who are neither wholly good nor wholly bad. It is much easier to identify the good from the bad in Heiss' novel, though. There are the bad settlers who, despite being humanised somewhat through their traumatic experience of the flood, know little to nothing about water and exploit Wagadhaany. Even the more ethically minded Louisa is, by the end of the novel, nothing more than a naïve and largely self-interested housewife.

While the above may seem like a limitation of Heiss' novel, it is in fact what makes it a particularly powerful work of historical fiction. The historical fiction genre is quite different to other genre fiction in that it perhaps exhibits less identifiable tropes and archetypes. It can therefore prove difficult to analyse in comparison to, say, crime fiction, that often follows a repeatable formula. Returning to Shaw's point from earlier, the historical novel largely does not deviate from "the formal techniques and cultural assumptions of the main traditions of the novel" (1983, 23). What is key here, though, is a self-awareness in Heiss' novel of the colonial history of the Australian historical novel (Birns 2018) and its use of this form to challenge the supposedly "established" and "true" colonial histories and ideologies that led to the development of the form in the first place. As Heiss herself writes, "I felt compelled to write [Yarri's] story into the national literary landscape to reach the broadest possible audience" (2022, 382).

In Heiss' historical note, she acknowledges various sources she has drawn from, many of these being colonial sources such as the *Goulburn Herald* (Heiss 2015, 384), the Aborigines Protection Board (385), Quakers Australia (386) and the *Masters and Servants Act 1840 No. 28a* (387). Following this with her author's note and acknowledgements,

Heiss recognises that her novel is equally "filled with the love, the knowledge, the wisdom, the support and the yindyamarra [respect] for Wiradyuri life and values of so many others" (2022, 391) as well as the knowledge and experience she gained during Wiradyuri language classes in Wagga Wagga at Charles Sturt University and "along the Marrambidya Bila" (Heiss 2022, 390). In bringing together these different forms of historical and cultural knowledge, Heiss' work reflects an entangled history involving the river, the Wiradyuri people and white settlers—a history that draws from multiple knowledges and stories rather than asserting a singular colonial "truth" about the Gundagai flood and its legacy.

Some events in the novel similarly critique the assumed superiority of colonial systems of knowledge production and dissemination. When Louisa insists that Wagadhaany introduce her to the local Wiradyuri women who live by the river in Wagga Wagga, Louisa expresses her desire to bring some books with her so that she can "teach some of the children how to read" (Heiss 2022, 196). Wagadhaany feels "deflated" (Heiss 2022, 196) by this, particularly when Louisa pulls out her bible with the intention of using this to teach them. Wagadhaany gently suggests that Louisa instead bring along "some of the books with pictures" next time, which she believes will be "better for the children" than the bible (Heiss 2022, 197). While Louisa agrees, Wagadhaany reminds herself that the books do not represent Wiradyuri culture—that the children will not be reading about or seeing pictures of "children fishing by the river, hunting, weaving and dancing" (Heiss 2022, 197). This passage reveals the inadequacy of a one-story-fits-all approach to history, challenging the idea of a singular narrative of place and people that is often enforced by colonial powers. To Louisa, a white settler woman convinced by the correctness of her form of knowledge, the ideas and stories in her picture books are culturally encompassing and therefore as applicable to the Wiradyuri people as they are to her own people. As she proclaims to the Wiradyuri women, the books "have pictures of children playing and having fun, just like your children" (Heiss 2022, 197). In reality, though, the people and the river in her picture books are just one story of many; Heiss' novel reinforces not only the inadequacy of this kind of storytelling, but the dangers of it in the face of ecological crisis.

7.3 Conclusion

Heiss' *Bila Yarrudhanggalangdhuray* is, of course, more than a fictional telling of the floods of Gundagai—it is an assertion of Wiradyuri sovereignty and the importance of kinship with Country in a volatile world, whether this be in the time of violent dispossession and oppression of the Wiradyuri people and other First Nations people in colonial-era Australia, or in the more contemporary time of climate crisis where this dispossession and oppression is still ongoing. Heiss' use of the historical novel represents an unsettlement of both the literary form itself and the human/nonhuman, culture/nature binaries that have ideologically informed it.

The novel's recognition of waterway agency is vastly different to the instrumental ways in which water and rivers have traditionally been represented in Australian historical fiction. Through this recognition and Heiss' use of a variety of sources to inform her story, the novel reveals both the short and long term consequences of an absence of relational understanding and kinship with the nonhuman amongst white settlers. Importantly, this novel is not just a story of the past but, as all historical novels are, a product of its present, emphasising the urgent need to challenge the still-dominant colonial ideologies of utilitarianism that inform much of Australia's waterway management today.

References

Bartha-Mitchell, Kathrin. 2024. *Cosmological Readings of Contemporary Australian Literature: Unsettling the Anthropocene*. London: Routledge.

Bila Yarrudhanggalangdhuray River of Dreams By Anita Heiss. 2021. *Simon & Schuster*. https://www.simonandschuster.com.au/books/Bila-Yarrudhanggalangdhuray/Anita-Heiss/9781761104800. Accessed 6 Jan 2025.

Birch, Tony. 2018. Friday Essay: Recovering a Narrative of Place - Stories in the Time of Climate Change. *The Conversation*, April 27. https://theconversation.com/friday-essay-recovering-a-narrative-of-place-stories-in-the-time-of-climate-change-95067. Accessed 6 Feb 2025.

Birns, Nicholas. 2018. The New Historical Novel: Putting Mid-Twentieth-Century Australia Into Perspective. *Commonwealth Essays and Studies* 41 (1): 7–18.

Carey, Peter. 2000. *True History of the Kelly Gang*. Brisbane: University of Queensland Press.

Cohen, Jeffrey Jerome. 2017. Drown. In *Veer Ecology: A Companion for Environmental Thinking*, ed. Jeffrey Jerome Cohen and Lowell Duckert, 246–267. Minneapolis: University of Minnesota Press.
Fetherston, Rachel. 2023. Unsettlement, Climate, and Rural/Urban Place-Making in Australian Crime Fiction. In *The Routledge Handbook of Crime Fiction and Ecology*, ed. Nathan Ashman, 78–90. New York: Routledge.
Flanagan, Richard. 2001. *Gould's Book of Fish: A Novel in Twelve Fish*. London: Picador.
Flanagan, Richard. 2013. *The Narrow Road to the Deep North*. New York: Knopf.
Gammage, Bill. 1983. The Wiradjuri War, 1838-40. *The Push* 16: 3–17.
Grenville, Kate. 2005. *The Secret River*. Melbourne: Text Publishing.
Griffiths, Tom. 2015. The Intriguing Dance of History and Fiction. *TEXT* 19 (Special 28): 1–21.
Groce, Eric, and Robin Groce. 2005. Authenticating Historical Fiction: Rationale and Process. *Education Research and Perspectives* 32 (1): 99–119.
Haraway, Donna. 2016. *Staying with the Trouble: Making Kin in the Chthulucene*. Durham: Duke University Press.
Heidegger, Martin. 1993. The Question Concerning Technology. In *Basic Writings*. London: Routledge.
Heiss, Anita. 2022 [2021]. *Bila Yarrudhanggalangdhuray River of Dreams*. Sydney: Simon & Schuster.
Janson, Julie. 2022. *Madukka The River Serpent*. Crawley, WA: UWA Publishing.
Jerome, John. 1997. *Blue Rooms: Ripples, Rivers, Pools, and Other Waters*. New York: Henry Holt.
Kent, Hannah. 2013. *Burial Rites*. Sydney: Pan Macmillan.
Kimmerer, Robin Wall. 2020 [2013]. *Braiding Sweetgrass: Indigenous Wisdom, Scientific Knowledge and the Teachings of Plants*. London: Penguin Books.
Lucashenko, Melissa. 2018. *Too Much Lip*. St Lucia, QLD: University of Queensland Press.
Lucashenko, Melissa. 2023. *Edenglassie*. St Lucia, QLD: University of Queensland Press.
McCullough, Colleen. 2000. *Morgan's Run*. New York: Simon & Schuster.
McCullough, Colleen. 2003. *The Touch*. New York: Simon & Schuster.
National Museum Australia (NMA). *Gundagai Flood*. https://www.nma.gov.au/defining-moments/resources/gundagai-flood-1852. Accessed 6 Feb 2025.
Neimanis, Astrida. 2016. *Bodies of Water: Posthuman Feminist Phenomenology*. London: Bloomsbury Academic.
O'Gorman, Emily. 2012. *Flood Country: An Environmental History of the Murray-Darling Basin*. Collingwood, VIC: CSIRO Publishing.
Potter, Emily. 2019. *Writing Belonging at the Millenium: Notes from the Field on Settler-Colonial Place*. Bristol: Intellect.

Riwoe, Mirandi. 2020. *Stone Sky Mountain*. St Lucia, QLD: University of Queensland Press.
Shaw, Harry E. 1983. *The Forms of Historical Fiction: Sir Walter Scott and His Successors*. Ithaca and London: Cornell University Press.
Shewry, Teresa, and Philip Steer. 2024. Uncertain Waters and Irony in Australian Settler Literatures. *Environmental Humanities* 16 (2): 351–370.
Simons, Margaret. 2020. The Tragedy of the Murray-Darling Basin. *Quarterly Essay* 7: 1–114.
Silcox, Beejay. 2021. *Bila Yarrudhanggalangdhuray* by Anita Heiss review – A mighty and generous heartsong. Guardian Australia, May 21. https://www.theguardian.com/books/2021/may/21/bila-yarrudhanggalangdhuray-by-anita-heiss-review-a-mighty-and-generous-heartsong. Accessed 6 Jan 2025.
Smith, Mick. 2017. Rain. In *Veer Ecology: A Companion for Environmental Thinking*, ed. Jeffrey Jerome Cohen and Lowell Duckert, 230–245. Minneapolis: University of Minnesota Press.
Soerjohardjo, Wardiningsih. 1986. *Aspects of Life in Gundagai, 1840–1860*. Master's thesis, The Australian National University, Australia.
Soerjohardjo, Wardiningsih. 2012. Remembering Yarrie: An Indigenous Australian and the 1852 Gundagai Flood. *Public History Review* 19: 120–129.
Sullivan, Jane. 2021. River of Dreams: Anita Heiss's New Novel. *Australian Book Review* 432. https://www.australianbookreview.com.au/abr-online/archive/2021/june-2021-no-432/964-june-2021-no-432/7820-jane-sullivan-reviews-bila-yarrudhanggalangdhuray-by-anita-heiss. Accessed Jan 6 2025.
Thrush, Coll. 2017. Haunt. In *Veer ecology: A companion for environmental thinking*, ed. Jeffrey Jerome Cohen and Lowell Duckert, 268–281. Minneapolis: University of Minnesota Press.
Wagner, Erica. 2021. *The Yield* by Tara June Winch review—Reclaiming Australia's indigenous voices. *Guardian Australia*, January 23. https://www.theguardian.com/books/2021/jan/23/the-yield-by-tara-june-winch-review-reclaiming-australias-indigenous-voices. Accessed 14 Jan 2025.
Wardle, Deborah. 2024. Understanding Aquifers Through Groundwater Stories: An Introduction. In *Understanding Aquifers Through Groundwater Stories*, ed. Deborah Wardle, 1–11. Deborah Wardle Books.
Weaver-Hightower, Rebecca. 2010. The Sorry Novels: Peter Carey's *Oscar and Lucinda*, Greg Matthews' *The Wisdom of Stones*, and Kate Grenville's *The Secret River*. In *Postcolonial Issues in Australian Literature*, ed. Nathanael O'Reilly, 129–152. Amherst: Cambria Press.
Winch, Tara June. 2019. *The Yield*. London: Hamish Hamilton.
Wright, Alexis. 2013. *The Swan Book*. Artarmon: Giramondo Publishing Company.

CHAPTER 8

Eco-literary Futures

8.1 Australian Ecofiction and the State of the Eco-novel

Arnaud Barras writes that "The world is storied, and being able to identify, interpret and understand its stories is a skill that matters" (2015, 1). This idea also applies to the natural world (if we are even to treat it separately from the human world in the first place): it is also a space that is storied, full of both human and nonhuman experiences that often defy interpretation but that can demand it nonetheless. If humans are to better understand their connection to and impact on the natural world, then an understanding of nature's stories is vital. While the narratives I have discussed throughout this book have been told using human words, they still reveal important stories of the nonhuman. Readers of these novels are privy to the experiences and, in some cases, thoughts and feelings of the nonhumans represented, and are exposed to nuanced understandings of what it means to connect with nature and the nonhuman in Australia. While it is no new claim that storytelling has the power to transport the reader into other worlds and situations, my hope is that this book has shed new light on how Australian ecofiction authors are representing the nonhuman in their works, and what such representations demonstrate to readers about serious ecological issues that are already impacting humans and

R. Fetherston, *Theorising the Postcolonial Eco-Novel*, Literatures, Cultures, and the Environment,
https://doi.org/10.1007/978-3-032-04466-2_8

nonhumans across the world. There is still much more to know about the kind of persuasive power that such narratives possess. However, I am hopeful that my work has paved the way just a little further to broaden the knowledge of scholars, in ecocriticism, the environmental humanities and beyond, of the role of literary narratives in environmental discourse and why the nonhuman stories within these narratives matter. Understanding the complexities of these stories is even more pertinent considering the advent of the COVID-19 pandemic and the global rise of far-right movements alongside the climate emergency, all of which have resulted in further pressure on humans' relationships with nature and the nonhuman as nations and communities struggle to keep overlapping health, political and environmental crises under control.

The postcolonial eco-novel has the potential to challenge outmoded preconceptions of nature and the nonhuman as conquerable entities that exist separately, and sometimes in opposition, to the settler-coloniser. It is of course not just the "what" or "who" of the nonhuman that matters (such as animals, plants, habitats, climate), but also the "how" of nonhuman representation in ecofiction. While many white, settler-Australian authors are still reliant on settler-colonial stereotypes regarding nature, some of the settler authors discussed in this book engage with more nuanced ways of perceiving the nonhuman that do not always advance the problematic Western, settler-colonial dualisms of culture/nature and human/nonhuman that persist in dominant Australian political and environmental discourse. Of course, First Nations authors such as Alexis Wright and Anita Heiss offer very different views of nature and the nonhuman, and their works play a vital role in revealing the agency of the Australian nonhuman as well as their plight in the face of colonisation and climate change. In different ways, Merlinda Bobis and Yumna Kassab critique certain aspects of white, colonial ideologies of belonging and human–land relations that in other texts, such as the crime novels, are largely endorsed.

The postcolonial eco-novel's concern with environmental crises and the continuing legacy of colonialism makes this form well-placed to interrogate the complexity of human-nonhuman relations in the twenty-first century. While some of the novels explored in this book refer clearly to anthropogenic climate change and associated environmental devastation, it is notable that others, such as the crime novels, do not explicitly mention the climate crisis. Rather, these texts allude to ecological crises like climate change through their depiction of bushfire and drought. The inclusion of popular genre texts such as these in ecocritical research can provide a broader understanding of Australian ecofiction as something that extends beyond "serious" literature, and into the realm of more broadly accessible fiction.

The nonhuman is a multifaceted concept that encapsulates animals, plants, habitats and broader ecosystems as well as climate, bushfire, drought, flood and the abiotic; any one example of a postcolonial eco-novel may attend to one or more of these conceptualisations in the context of (post)colonialism and the Anthropocene. These nonhumans may be represented as individuals, such as the fig tree in Catherine McKinnon's *Storyland* (2017), and as whole species, such as the black swan in Wright's *The Swan Book* (2013) and the Australian giant cuttlefish in Jennifer Mills' *Dyschronia* (2018). They may be grouped together within habitats, a nonhuman entity in itself; for example, the many nonhumans referred to throughout *Storyland* that inhabit the Lake Illawarra area present an understanding of the nonhuman as both within and of habitat. Additionally, there are droughts, bushfires and floods that are a part of Australia's natural climate cycles but that are also strongly influenced by anthropogenic climate change (Jones et al. 2013), plus the abiotic nonhuman that may cover "natural" objects, such as rocks, or human-made objects, such as farming equipment and furniture. This expansive definition of the nonhuman is one that can and should be applied to future analyses of ecofiction works, as it refuses a homogenisation of these organisms, objects and phenomena, instead providing space to ask how and why the complexity of the nonhuman opens up more inclusive ways of understanding the far-reaching consequences of environmental crisis and its origins in the colonial project.

I have discussed how some Australian ecofiction—such as McKinnon's and Mireille Juchau's *The World Without Us* (2015)—challenges certain settler-colonial stereotypes about the nonhuman that are often employed in Australian settler literature, whilst also still working within this settler-colonial literary tradition. Juchau's novel in particular succumbs to some problematic tropes that present habitat as a more representational force in the narrative than as a character in its own right. Nevertheless, these texts address the present and future impacts of climate change on Australian habitats and the role that colonisation has played, and is still playing, in ecological crises. Taking inspiration from Libby Robin's claim that habitats are "nonhuman actors" within Australia's evolving national identity (2007, 2), I believe that habitat stories in particular can deeply interrogate connections between non-Indigenous Australians and the nonhuman, and reveal the possibilities of applying more diverse understandings of the Australian nonhuman to the climate crisis. This kind of settler-authored novel, while not entirely unproblematic, indicates a shift away from the largely symbolic use of the nonhuman that has been prevalent in contemporary Australian literature for so long, and a shift towards a more relational connection between the (white settler) human and the nonhuman in the midst of climate catastrophe.

Some of the novels I have looked at, primarily the crime novels, still rely heavily on the symbolism of nonhumans rather than emphasising the material nonhuman and their experiences of drought alongside the human. Additionally, these works reproduce a colonial outlook on the Australian natural world, with a focus on Anglo-agricultural landscapes and non-native farm animals that are arguably a primary reason for much of the devastation caused to native habitats and species across the country. Nevertheless, the authors' exploration of the links between nonhuman and human communities, whilst symbolic to an extent, also reflects an important cause-and-effect relationship. As the climate crisis intensifies, nonhumans will continue to suffer and die alongside the humans affected; the depiction of how this occurs in Chris Hammer's (2018) and Jane Harper's *The Dry* (2016) is therefore not simply symbolic of human suffering alone, but the very real connection between the suffering of both humans and nonhumans under climate change.

So, what does the future of the postcolonial eco-novel look like? While it is true that we are seeing an upward trend in the publication of Australian environmental fiction, particularly climate fiction, it is also true that conceptions of the nonhuman in the context of the climate crisis and the Anthropocene are in constant flux. An increased focus on popular genre in this respect may reveal some interesting possibilities regarding the broader communication of climate crisis and colonial culpability beyond the realm of the literary highbrow, both in Australia and internationally.

While in the recent past scholars like Amitav Ghosh have maligned the relegation of climate fiction to the science fiction genre (2016, 7), I would argue that, as well as publishing more literary climate fiction, literary communities should continue to work towards a broader acceptance of genre fiction, like science fiction, crime fiction, fantasy fiction and romance fiction, into serious literary spaces. This is because, as I have shown in this book, the shift to more nuanced representations of the nonhuman and an increased concern with the impacts of climate change, colonialism and capitalism extends to genre fiction, making it worthy of consideration in this respect. The absence of direct references to the climate crisis in the selected crime texts means that such novels do not carry an overt environmental message in comparison to the more explicit environmental messages of the other, more literary works discussed. This might seem problematic to those writing and researching more serious environmental literature, but for those wanting to connect with a broader audience through pro-environmental messaging this may, in fact, be an advantage—these texts are potentially introducing different groups of readers to environmental themes in a more accessible way.

There is a great need for environmentalists, such as activists and scientists, to demonstrate to the general public the links between perceived "natural" disasters, anthropogenic climate change and colonialism, and yet there are still various avenues for doing so that remain largely unexplored. In his work on climate fiction, Matthew Schneider-Mayerson draws on climatologist Judith Curry who comments that while "scientists and other people are trying to get their message across about various aspects of the climate change issue … fiction is an untapped way of doing this—a way of smuggling some serious topics into the consciousness of readers" (Evancie 2013, cited in Schneider-Mayerson 2018, 475). While I hesitate to apply an entirely empirical value to literature in this sense, it is certainly an exciting possibility that crime fiction, amongst other genre fiction, may be well placed to convey the threat of climate change and other environmental crises to a broader populace. I do not raise this point with the intention of denigrating the value of more hermeneutical ecocritical research. Nevertheless, I believe that extending ecocritical and postcolonial scholarship to focus more so on genre fictions is an important step in understanding the current crisis of communication and ideology surrounding climate change and the enduring legacy of colonialism.

As we are also seeing the publication of more Indigenous Australian authors, it is likely that Australian ecofiction will continue to encapsulate works like Wright's *The Swan Book* and Anita Heiss' *Bila Yarrudhanggalangdhuray* (2021) that engage with Indigenous understandings of the nonhuman, particularly in the context of how Indigenous genocide and dispossession at the hands of settler-colonisers have impacted

Australia's lands and waters, and how the climate crisis specifically affects Indigenous Australian people in different ways to settlers. It must also be noted that the more complex depictions of the nonhuman by settler-colonial Australian authors—the ones that challenge Western conceptions of Australian nature—often draw from Indigenous Australian ontologies and epistemologies, including First Nations storytelling practices employed in contemporary Aboriginal literature. As a result, it is important that scholars recognise how settler authors are strongly influenced by Indigenous knowledges and storytelling regarding the nonhuman, rather than make claims that non-Indigenous authors are creating entirely new representations of nature and the nonhuman through their narratives.

8.2 Ecocriticism and Decolonial Futures

In Chap. 3, I discussed how the protagonists' ability to converse with the nonhuman in Wright's *The Swan Book* and Bobis' *Locust Girl* (2015) is representative of the human relationship with these nonhumans, but also provides a means of understanding the nonhuman experience itself. Both novels break down the traditional and largely Western dualisms of culture versus nature and human versus animal through their depiction of multispecies relationships from the perspective of both humans and nonhumans. Connecting with the nonhuman other is portrayed by Wright and Bobis as a positive, yet complex, means by which to overcome the boundaries between human and nonhuman difference. The strong bonds formed between each novel's protagonist and the nonhuman animal demonstrate a more relational approach to some of the environmental issues specific to Australia and other postcolonial nation states where colonial notions of borders and property enclosure still assert the separation of human/nonhuman, culture/nature and coloniser/colonised.

A possible remedy suggested by Wright and Bobis is a closer relationship with the nonhuman beings who share our homes—a concept that is central to the Indigenous Australian understanding of Country, as well as in multispecies studies that draws on this and other Indigenous ontologies and epistemologies (van Dooren et al. 2016; van Dooren and Rose 2016; Ogden et al. 2013; Haraway 2016). These acts of "kin-making", as Haraway (2015) describes, are a means of subverting current settler-colonial paradigms regarding the natural world that have contributed to the climate emergency. In Wright's work especially, Indigenous dispossession is intimately linked to her depiction of climate catastrophe. Her emphasis on Oblivia's "sovereignty of the mind" (Wright and Zable 2013) is more broadly representative of Indigenous sovereignty of Country, while the relationship Oblivia develops with the black swans, both in her mind and in the world around her, suggests that this is part of her deeper relationship with the land. This relationship reveals the novel as part of a broader decolonising literary practice and demonstrates an imperative for ecocritical research to be more actively engaged with decolonial literary theory and, in some cases, real-world decolonising practices outside of literature.

Some of the most vital research yet to be done in an Australian context is how narratives in general can impact audience environmental attitudes, values and behaviours, particularly in the context of the climate crisis and how this links to bushfires, drought, and flooding across the nation. While researchers have undertaken studies that investigate a similar question outside of Australia (see Schneider-Mayerson 2018, 2020; Malecki et al. 2019), little is known about how Australian audiences, such as readers, respond to environmental narratives. While this book has not specifically explored this question, there is an imperative to better comprehend how different textual forms represent the links between colonialism and environmental crisis in affective ways for readers. Can literature actually lead to decolonial or environmental action?

My research provides a starting point for this, with a hope that increased understandings of how such narratives impact readers may enable activists, writers, artists and communicators across various media to engage with nuanced environmental perspectives more strongly, particularly regarding the nonhuman. A focus on reader or audience emotion and attitude may be relevant to these queries. My own reader response study, which formed a part of this original research project and was co-developed with Emily Potter, Kelly Miller and Devin Bowles, provides some insight (Fetherston et al. 2023), but a more expansive approach is required to understand differences in representation across both fiction and non-fiction forms.

I wrote this book in the hope of better understanding how the nonhuman factors into Australian ecofiction narratives and how the nonhuman has the ability to destabilise, or unsettle, settler-colonial belonging. Since colonisation, Australian native species and habitats have faced immense threats, while serious and effective action on climate change is severely lacking at a federal government level in Australia today. It is already clear to many Australians—scientists, artists, writers, academics, activists, students and more—that something has to change. With what is already known about the persuasive power of narrative, storytelling may be one means of communicating environmental crisis to a wider range of people, especially voters. While this may seem like wishful thinking (and my research here has certainly not proven this beyond any reasonable doubt), there is the possibility that the act of reading ecofiction and being influenced by nonhuman narratives can manifest into more people encouraging decision-makers in both Australia and elsewhere to play a more impactful role in both national and international climate negotiations. As Potter concludes, "new imaginaries are needed that break from a hermetic, colonial vision for non-indigenous narratives" (2019, 147). Such imaginaries should be more attuned to the nuances of human–land relations for both Indigenous and non-Indigenous people if they are to unsettle the dominant Western-colonial binaries of human/nonhuman, culture/nature, coloniser/colonised and subject/object, to name but a few.

Subsequently, the way Australians think, write and read about nature and the nonhuman matters. Living beings, human and nonhuman alike, live and die based on how their stories are told. As much as it may seem removed from the real world of climate action and politics, imbuing the nonhuman with sentience, agency and purpose through fiction, beyond the purely symbolic or metaphoric, is not an action to be taken lightly—stories can recognise the agency, personhood and complex life-worlds of the more-than-human, or they can deny them. Both kinds of stories have consequences for how we too, as humans, live and act in a world that remains dominated by colonial-capitalist hierarchies of power.

While fiction alone may not be the solution, it is fundamental that academics across disciplines continue to acknowledge the damaging role that settler-colonial ideologies play in contemporary environmental discourse and policy across the world. Recognising that Indigenous Australian peoples, amongst many other Indigenous peoples globally, have long embraced deeply relational considerations of nonhuman experience and agency is just as vital. It is therefore important that we understand the damage done by the settler-colonial project in Australia and beyond while challenging the project's dominance via more ecocentric considerations of both human and nonhuman experiences.

References

Barras, Arnaud. 2015. The Law of Storytelling: The Hermeneutics of Relationality in Alexis Wright's *The Swan Book*. *Journal of the Association for the Study of Australian Literature* 15 (3): 1–12.

Bobis, Merlinda. 2015. *Locust Girl: A Lovesong*. North Melbourne, Spinifex Press.

Evancie, Angela. 2013. So Hot Right Now: Has Climate Change Created a New Literary Genre? *National Public Radio*, April 20. http://www.npr.org/2013/04/20/176713022/so-hot-right-now-has-climate-change-created-a-new-literary-genre. Accessed 7 Feb 2025.

Fetherston, Rachel, Emily Potter, Kelly Miller, and Devin Bowles. 2023. Seeking Greener Pages: An Analysis of Reader Response to Australian Eco-Crime Fiction. *Australian Humanities Review* 71 (May): 1–21.

Ghosh, Amitav. 2016. *The Great Derangement: Climate Change and the Unthinkable*. Chicago and London: University of Chicago Press.

Hammer, Chris. 2018. *Scrublands*. Crows Nest: Allen & Unwin.

Haraway, Donna. 2015. Anthropocene, Capitalocene, Plantationocene, Chthulucene: Making Kin. *Environmental Humanities* 6 (1): 159–165.

Haraway, Donna. 2016. *Staying with the Trouble: Making Kin in the Chthulucene*. Durham: Duke University Press.

Harper, Jane. 2016. *The Dry*. Sydney: Pan Macmillan Australia.

Jones, Roger, Celeste Young, John Handmer, Adrian Keating, Gayathri Mekala, and Peter Sheehan. 2013. *Valuing Adaptation Under Rapid Change*. Gold Coast, QLD: National Climate Change Adaptation Research Facility.

Juchau, Mireille. 2015. *The World Without Us*. London: Bloomsbury Publishing.

Malecki, Wojciech, Piotr Sorokowski, Boguslaw Pawlowski, and Marcin Cieński. 2019. *Human Minds and Animal Stories: How Narratives Make Us Care About Other Species*. New York and London: Routledge.

McKinnon, Catherine. 2017. *Storyland*. Sydney: HarperCollins Publishers.

Mills, Jennifer. 2018. *Dyschronia*. Sydney: Picador.

Ogden, Laura A., Billy Hall, and Kimiko Tanita. 2013. Animals, Plants, People, and Things: A Review of Multispecies Ethnography. *Environment and Society* 4: 5–24.

Potter, Emily. 2019. *Writing Belonging at the millennium: Notes from the Field on Settler-Colonial Place*. Bristol: Intellect Books.

Robin, Libby. 2007. *How a Continent Created a Nation*. Sydney: University of New South Wales Press.

Schneider-Mayerson, Matthew. 2018. The Influence of Climate Fiction: An Empirical Survey of Readers. *Environmental Humanities* 10 (2): 473–500.

Schneider-Mayerson, Matthew. 2020. "Just as in the Book"? The Influence of Literature on Readers' Awareness of Climate Justice and Perception of Climate

Migrants. *Interdisciplinary Studies in Literature and Environment* 27 (2): 337–364.

Van Dooren, Thom, and Deborah Bird Rose. 2016. Lively Ethography: Storying Animist Worlds. *Environmental Humanities* 8 (1): 77–94.

Van Dooren, Thom, Eben Kirksey, and Ursula Münster. 2016. Multispecies Studies: Cultivating Arts of Attentiveness. *Environmental Humanities* 8 (1): 1–23.

Wright, Alexis. 2013. *The Swan Book*. Artarmon: Giramondo Publishing Company.

Wright, Alexis, and Arnold Zable. 2013. The Future of Swans. *Overland Literary Journal* 213 (Summer). https://overland.org.au/previous-issues/issue-213/feature-alexis-wright-and-arnold-zable/

Index[1]

[1] Note: Page numbers followed by 'n' refer to notes.

R. Fetherston, *Theorising the Postcolonial Eco-Novel*, Literatures, Cultures, and the Environment,
https://doi.org/10.1007/978-3-032-04466-2

Kolibri 360 GmbH
Ferdinand-Jühlke-Straße 7
99095 Erfurt, Deutschland
produktsicherheit@kolibri360.de